# A Sequential Introduction to
# Real Analysis

# Essential Textbooks In Mathematics

ISSN: 2059-7657

Essential Textbooks in Mathematics

# A Sequential Introduction to
# Real Analysis

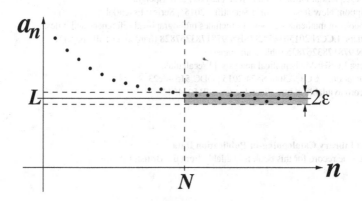

## J M Speight
University of Leeds, UK

 **World Scientific**   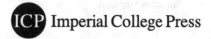 **Imperial College Press**

*Published by*

World Scientific Publishing Europe Ltd.

57 Shelton Street, Covent Garden, London WC2H 9HE

*Head office:* 5 Toh Tuck Link, Singapore 596224

*USA office:* 27 Warren Street, Suite 401-402, Hackensack, NJ 07601

**Library of Congress Cataloging-in-Publication Data**
Names: Speight, J. M. (J. Martin)
Title: A sequential introduction to real analysis / J.M. Speight.
Description: New Jersey : World Scientific, 2015. | Series: Essential
   textbooks in mathematics ; vol. 1 | Includes bibliographical references and index.
Identifiers: LCCN 2015034542| ISBN 9781783267828 (hardcover : alk. paper) |
   ISBN 9781783267835 (pbk. : alk. paper)
Subjects: LCSH: Mathematical analysis. | Precalculus.
Classification: LCC QA300 .S874 2015 | DDC 515--dc23
LC record available at http://lccn.loc.gov/2015034542

**British Library Cataloguing-in-Publication Data**
A catalogue record for this book is available from the British Library.

Printed in Singapore

*To my parents, Doreen and Dennis, with love and gratitude (because the Town Moor beats the Top End any day).*

# Preface

It is a truth universally acknowledged that no university worth its salt should turn out mathematics graduates who don't understand calculus properly. Real analysis remains, therefore, a core component of most university mathematics curricula. Traditionally, undergraduate students have found real analysis rather challenging, and this has not noticeably improved in recent years, despite (in the UK) our incoming students having enjoyed an unprecedented regime of consistent and rigorous testing at school. One suspects that the drive to "teach to the test", with its emphasis on method over meaning, might even be counterproductive. So what is it that makes real analysis difficult? The underlying mathematical concepts are not that subtle: for the most part, real analysis concerns nothing more abstract than the idea of one number being bigger than another. The subtlety lies in the thick accumulation of *quantifiers* that quickly builds up in any definition or argument.

It seems to me that, in this regard, the notion of the limit of a convergent *sequence* is intrinsically easier than the limit of a function. Both concern two variables and are of the form "for all values of variable 1, there exists a value of variable 2", but in the case of sequences, the variables are of entirely different natures: one is a (small) positive real number while the other is a (large) positive integer, and this distinction is reinforced by the names we commonly give them ($\varepsilon$ and $N$). They are conceptually quite distinct, and there is little danger of confusing them. Indeed, there is only one possible meaning of the phrase "limit of a sequence". This makes the structure of a typical sequential convergence proof ("for each $\varepsilon > 0$ we must find a positive integer $N$ such that ...") rather plain and clear. By contrast, the variables occuring in the limit of a function are both (small) positive real numbers, typically given similar names to one another ($\varepsilon$ and

$\delta$), and proofs in this mould can easily throw up a bewildering array of
small numbers whose inter-relationship is difficult to apprehend. There
are several possible limits, one for each cluster point of the domain (and
even defining cluster points is nontrivial), not to mention one-sided limits
and other elaborations. The traditional approach to real analysis is to treat
convergence of sequences quickly, as warm-up material, before moving on to
the more serious $\varepsilon$–$\delta$ stuff. The hope is that, having been exposed (briefly)
to the subtlety engendered by quantifiers in the simpler setting of sequences,
students will be able to cope in the more challenging context of functions of
a real variable. Experience suggests that this hope is very often misplaced.

This book is based on the lecture notes for a first year course, taught at
the University of Leeds, that takes a somewhat different tack.[1] We spend
a lot of time studying convergence of real sequences in great detail, until
the $\varepsilon$–$N$ notion of convergence is thoroughly ingrained. Then, rather than
shifting to $\varepsilon$–$\delta$ notions, we use sequences and their limits to *define* continuity
and limits of functions. In this way, the part of real analysis that students
typically struggle with (the quantifiers concerning pairs of large and/or
small quantities) is localized and compartmentalized. Having thoroughly
understood convergence of sequences, it is not so hard to formulate elegant
sequential definitions of continuity and limit. These can be exploited to
give a rigorous development of calculus, including all the classical theorems
one expects to encounter (the Intermediate, Extreme, and Mean Value
Theorems, Taylor's Theorem, the Fundamental Theorem of the Calculus
etc.). This, to my mind, is the fundamental goal of undergraduate real
analysis, and I have ruthlessly stripped out everything extraneous to its
attainment: not just all $\varepsilon$–$\delta$ criteria, including the notion of uniform
continuity, but even the Cauchy criterion for convergence of a sequence
(though this does appear in Chapter 13, where we construct the reals as
the completion of the rationals).

The intended audience for this book is first and second year
undergraduate students of mathematics, though students of other scientific
disciplines with a strong mathematical inclination may also profit from it.
It assumes working familiarity with the basic notions of calculus, at an
informal level, and with the technique of proof by induction. It contains a
chapter on symbolic logic, including quantifiers. Note that this is *not* an
appendix: it seems to me that a first real analysis course is exactly the right
place to introduce students to this material. It is placed, quite deliberately,

---

[1]The course concerned covers Chapters 1 to 8 and part of 10. The remaining material
is covered in the second year.

about half-way through the book so that students learn to formalize and algebraically manipulate complex statements only after they have gained significant experience in dealing with them through the medium of standard English. The book proper ends with a chapter in which we *construct* the field of reals as a space of equivalance classes of rational Cauchy sequences, a rather unusual inclusion in an introductory analysis text (Dedekind cuts being the more usual approach, if this aspect is covered at all) but one which fits into our sequential programme beautifully. This is somewhat advanced material for the core intended audience and is mainly for the benefit of extremely motivated students who want a glimpse of how deep the rabbit hole goes. I have tried to maintain throughout the informal, conversational tone of the lecture notes on which the book is based. I hope that any of my elders or betters who read it won't find this tone too jarring. Each chapter ends with a summary and two sets of exercises. The tutorial problems are intended to be studied in small groups with a tutor. Solutions to these may be found at the back of the book. The homework problems should be attempted alone. Of course, mathematics can only really be learned by *doing* it, so the exercises are a vital and integral part of the book. It is my intention to produce an electronic solution manual for these, which accredited academic faculty members can obtain from the publishers.

Finally, there is a pun in the title and, since I'm English, I'd better apologize for it. This is a *sequential* introduction to real analysis in the obvious sense: it consistently characterizes all the fundamental analytic concepts it uses in terms of sequences. But it is also sequential in another sense. Like any mature branch of mathematics, real analysis has been honed to form an elegant, interconnected mechanism, in which definitions and theorems build on one another. So, for example, the second version of the Fundamental Theorem of the Calculus relies on the Mean Value Theorem, which relies on the Interior Extremum Theorem, which relies on the Extreme Value Theorem, which relies on the Bolzano-Weierstrass Theorem, which relies on the Monotone Convergence Theorem, which relies on the Axiom of Completeness, and hence these results are presented in exactly the reverse of this order. In this sense, real analysis provides some of the most beautiful examples of mathematics' inherently sequential, cumulative nature.

Martin Speight
June 2015
Leeds, UK

# Foreword

## Why study real analysis?

Differential calculus is arguably the single most powerful and influential mathematical idea ever invented. Its applications pervade the natural sciences, and it is essential to most applied mathematics and much pure mathematics too (for example, in Differential Geometry, my own area of interest). The central object in differential calculus is the *derivative* of a function, and you have probably expended much intellectual effort in learning and becoming fluent in the techniques for computing derivatives.

But what, precisely, *is* a derivative? At school, it's common to be fobbed off with a vague "definition" like

"$f'(x)$ is the slope of the graph $y = f(x)$ at the point $(x, f(x))$".

But this isn't a definition at all because you don't have an independent definition of the word "slope". (The reality is that "slope" is defined by $f'(x)$, not the other way round!) Once you've completed an introductory calculus course at university, you know that the *real* definition is

$$f'(x) = \lim_{h \to 0} \frac{f(x+h) - f(x)}{h}.$$

But actually this isn't satisfactory either because it contains the mysterious symbol lim, that is, a *limit*. In fact, all the basic notions of calculus (derivatives, integrals, continuity) concern limits in one way or another. This merely begs the question, what does

$$\lim_{x \to a} g(x)$$

actually mean? In introductory calculus it is usual, again, to be fobbed off with something rather vague like

"$\lim_{x \to a} g(x) = L$ means that, as $x$ gets arbitrarily close to (but remains different from) $a$, $g(x)$ gets arbitrarily close to $L$".

This is certainly better than the "definition" of $f'$ in terms of "slope", but it still isn't quite satisfactory because the precise meaning of "arbitrarily close" isn't clear.

The main purpose of this book is to fill in this gap. We do so by introducing, and studying in considerable depth, the first and simplest rigorous definition of limit, namely, the limit of a convergent *sequence*. It turns out that limits of functions, and hence continuity and derivatives, can be defined purely in terms of sequential limits. So, by the end of Chapter 8 we will have developed a proper mathematical understanding of the basic object underlying calculus. In subsequent chapters we go on to use the sequential notion of limit to develop the theory of differential calculus and Riemann integration (rigorously proving the Mean Value Theorem and the Fundamental Theorem of the Calculus, for example). This development builds on some fundamental theorems about *continuous* functions, which we study in depth in Chapter 6, such as:

**The Intermediate Value Theorem** Let $f : [a, b] \to \mathbb{R}$ be continuous and $y$ be a number between $f(a)$ and $f(b)$. Then there exists $c \in [a, b]$ such that $f(c) = y$.

**The Extreme Value Theorem** Let $f : [a, b] \to \mathbb{R}$ be continuous. Then $f$ attains both a maximum and a minimum value on $[a, b]$.

Along the way we will also develop some tests and techniques for determining the convergence properties of sequences and series which are important and useful in their own right.

The endeavour of making limits rigorous is called "Real Analysis" by mathematicians. It is common for students, when first exposed to this subject, to ask

> "What's the point? All this stuff is obvious, why do we have to prove it?"

I have two responses to this. First, proving our assertions is what makes mathematics mathematics; it is the defining characteristic of our[2] discipline. Being able to construct and communicate a rigorous proof is a core skill which should be mastered by anyone hoping to graduate with a degree in mathematics, and an introductory course in real analysis is one of the first

---

[2] I speak here as someone whose first degree is in Physics, and hence with the zeal of a convert!

places where you will (hopefully) acquire this skill. Second, just because something is "obvious" doesn't mean it's true! Consider the following statements:

**Assertion 1** There is no continuous onto map from $\mathbb{R}$ to $\mathbb{C}$.

Such a map would define a continuous curve which fills the whole plane: could *you* draw such a curve – without taking your pencil off the paper?

**Assertion 2** If $f'(a) > 0$ then there is some neighbourhood of $a$ on which $f(x)$ is an increasing function.

The slope of the function is positive at $a$, so surely it must be going "uphill" as $x$ increases, at least for $x$ sufficiently close to $a$.

These are, to my mind, just as intuitively "obvious" as the Intermediate and Extreme Value Theorems. They are also **false!**[3]

---

[3]Disproving Assertion 1 is beyond the scope of this book, but you'll find a counterexample to Assertion 2 in Chapter 9.

# Acknowledgements

I am indebted to my colleagues in the School of Mathematics at the University of Leeds who have helped me teach the course on which this book is based: it has been improved considerably by their comments, suggestions, and objections. In particular, I'd like to thank Kevin Houston, who has thought long and hard about how to teach analysis. I've profited greatly from our discussions, and he will surely recognize his influence in what follows.

While the particular formulation of real analysis contained herein is my own work, it would, of course, be absurd to attach any claim of originality to this material. Even where I have given a (to the best of my knowledge) "new" sequential version of a known result, it is inconceivable, in such a time-honoured field, that I am the first person to discover it. The book's novelty lies in the consistent and distinctive sequential perspective it takes. A short list of further reading at the end of the book details the texts that I have found useful in learning real analysis myself, some texts that give a complementary view of the material covered here, and some which take the subject further.

Additional Resources for Instructors and Lecturers

A solutions manual for instructors and lecturers is available for download from the publisher's website: `http://www.worldscientific.com/worldscibooks/10.1142/p1032-sm`

Please note the download is passcode protected and that the code will not be issued to students. For authorised personnel, the code can be obtained from the World Scientific sales team.

# Contents

# Chapter 1

# Basic properties of the set of real numbers

## 1.1 Recap of set notation

In this first chapter, we introduce our main object of study, the set of real numbers, focussing on its crucial property of *completeness*. Here, and throughout the text, we make extensive use of the language of set theory, to which I assume the reader has had some prior exposure. We begin with a rapid recapitulation of its main points.

For our purposes, a **set** is a collection of objects, real or of the imagination. Given a set $A$, we write $x \in A$ to denote that object $x$ is in set $A$. This is usually translated into words (when reading the expression) as "$x$ is an **element** of $A$". The expression $x \notin A$ means $x$ is *not* an element of $A$.

Another set $B$ is a **subset** of $A$ if every element of $B$ is an element of $A$. This is denoted $B \subseteq A$. Note that every set is a subset of itself, $A \subseteq A$. Two sets $A$, $B$ are equal if they have precisely the same elements, in which case $A \subseteq B$ and $B \subseteq A$.

There are a few sets which occur frequently in mathematics and so have been given special, commonly agreed symbols.

$\emptyset$   The **empty set**. This is the set containing no objects. Do not confuse it with $\{\emptyset\}$, which certainly isn't the empty set since it contains an object (namely, the empty set)!

$\mathbb{N}$   The set of **natural numbers**, $\{0, 1, 2, 3, \ldots\}$.

$\mathbb{Z}$   The set of **integers**, $\{\ldots, -2, -1, 0, 1, 2, \ldots\}$.

$\mathbb{Z}^+$  The set of **positive integers**, $\{1, 2, 3, \ldots\}$. Note that 0 is neither positive nor negative.

$\mathbb{Q}$   The set of **rational numbers**, $\{\frac{a}{b} : a, b \in \mathbb{Z}, b \neq 0\}$.

$\mathbb{R}$   The set of **real numbers**.

This list illustrates the two most common ways of specifying a set, namely, listing its elements,

$$A = \{1, 2, 3, 4, 5, 6, 7, 8\},$$

and giving a condition that elements must satisfy,

$$A = \{x \; : \; x \in \mathbb{Z}^+, \, x \leq 8\}.$$

In general $A = \{x \; : \; P(x)\}$ means the set of objects $x$ for which condition $P(x)$ is true. If part of condition $P(x)$ is that $x$ should be in a particular set, we often move that part to the left of the colon. For example, the set above could be written

$$A = \{x \in \mathbb{Z}^+ \; : \; x \leq 8\}.$$

If there *is* no $x \in B$ such that $P(x)$ is true, then

$$A = \{x \in B \; : \; P(x)\} = \emptyset.$$

For example, $A = \{x \in \mathbb{Q} \; : \; x^2 = 2\} = \emptyset$, that is, there is no rational number whose square is 2. Let's prove it.

**Proposition 1.1.** *Let* $x \in \mathbb{Q}$*. Then* $x^2 \neq 2$.

**Proof.** Assume the claim is false and a rational number $x$ such that $x^2 = 2$ exists. Clearly $x \neq 0$. Since $(-x)^2 = x^2$, we can assume that $x > 0$. Hence there exist positive integers $p, q$ such that $x = p/q$, so

$$\frac{p^2}{q^2} = 2.$$

We may assume that $p, q$ have no common divisors (if they did, we could divide them all out to obtain new positive integers $p', q'$ with no common divisors such that $x = p'/q'$). Now $p^2 = 2q^2$, so $p^2$ is an even integer. It follows that $p$ must be an even integer (if it were odd then $p^2$ would be a product of odd integers and hence would itself be odd). That is, $p = 2k$ for some $k \in \mathbb{Z}^+$. But then $(2k)^2 = 2q^2$, that is, $q^2 = 2k^2$. Since $q^2$ is even, so is $q$. Hence $p, q$ have a common divisor (namely 2, since they are both even), which contradicts their definition. $\qquad\square$

Given real numbers $a, b$, with $a < b$, we define several different

**intervals**, as follows:

$$(a, b) = \{x \in \mathbb{R} \; : \; a < x < b\},$$
$$[a, b] = \{x \in \mathbb{R} \; : \; a \leq x \leq b\},$$
$$[a, b) = \{x \in \mathbb{R} \; : \; a \leq x < b\},$$
$$(a, b] = \{x \in \mathbb{R} \; : \; a < x \leq b\},$$
$$(a, \infty) = \{x \in \mathbb{R} \; : \; x > a\},$$
$$[a, \infty) = \{x \in \mathbb{R} \; : \; x \geq a\},$$
$$(-\infty, b) = \{x \in \mathbb{R} \; : \; x < b\},$$
$$(-\infty, b] = \{x \in \mathbb{R} \; : \; x \leq b\}.$$

So a round bracket denotes that the end point is omitted from the interval, while a square bracket denotes that it is included. Note that $\infty$ is *not* a number, so expressions like $(a, \infty]$ do not make sense. Intervals which *include* all their endpoints, that is, $[a, b]$, $[a, \infty)$, $(-\infty, b]$ are said to be **closed**. Those which exclude all their endpoints, that is, $(a, b)$, $(a, \infty)$, $(-\infty, b)$ are said to be **open**.

Given two sets $A$, $B$, there are several useful ways of constructing new sets from them.

- The **intersection** of $A$ and $B$ is

$$A \cap B = \{x \; : \; x \in A \text{ and } x \in B\}.$$

- The **union** of $A$ and $B$ is

$$A \cup B = \{x \; : \; x \in A \text{ or } x \in B\}.$$

Note that, in mathematics, we use "or" in the non-exclusive sense, that is, "$P$ or $Q$" means $P$ is true or $Q$ is true (or both).

- The **complement** of $B$ in $A$ is

$$A \backslash B = \{x \; : \; x \in A \text{ and } x \notin B\}.$$

- The **cartesian product** of $A$ and $B$ is

$$A \times B = \{(x, y) \; : \; x \in A \text{ and } y \in B\},$$

that is, the set of **ordered pairs** of objects $(x, y)$, where $x$ comes from $A$ and $y$ comes from $B$. Do not confuse the symbol $\times$, used in the context of sets, with multiplication.

**Example 1.2.**

$$\{1, 2, 3, 4, 5, 6, 7, 8\} = [1, 8] \cap \mathbb{Z}$$
$$= (-\infty, 9) \cap \mathbb{Z}^+$$
$$= \{1, 3, 5, 7\} \cup \{2, 4, 6, 8\},$$

$$\mathbb{R} \setminus \{-1, 1\} = (-\infty, -1) \cup (-1, 1) \cup (1, \infty),$$

$$\{\text{me}, \text{you}\} \times \{1, 2, 3\} = \{(\text{me}, 1), (\text{me}, 2), (\text{me}, 3), (\text{you}, 1), (\text{you}, 2), (\text{you}, 3)\},$$

$$(-2, \infty) \cap (-\infty, 0] = (-2, 0].$$

$\square$

## 1.2  Functions

Given two sets $A$, $B$, a **function** (or **mapping**) $f$ from $A$ to $B$ is a rule which assigns to each element of $A$ an element of $B$. The shorthand for this is $f : A \to B$, often read as "$f$ maps $A$ to $B$". The set $A$ is called the **domain** of $f$ and the set $B$ is called the **codomain** of $f$. For each $x \in A$, we denote by $f(x)$ the element in $B$ which the function $f$ assigns to the element $x$, and call this element the **image** of $x$ under $f$. There is no reason why two different elements in $A$ cannot have the same image under a function.

It is often helpful to think of $f$ as a machine which converts things in $A$ to things in $B$. Given an input $x \in A$, the machine spits out an output $f(x) \in B$. The set of all outputs is called the **range** of $f$, denoted $f(A)$. In set notation, the range is

$$f(A) = \{f(x) : x \in A\}.$$

By definition, this is a subset of the codomain $B$.

**Definition 1.3.**  A function $f : A \to B$ is **injective** (or **one-to-one**) if $f(x_1) = f(x_2)$ implies that $x_1 = x_2$. It is **surjective** (or **onto**) if $f(A) = B$. It is **bijective** if it is both injective and surjective.

We may restate these conditions as follows:

- $f$ is injective if whenever $x_1 \neq x_2$, $f(x_1) \neq f(x_2)$, that is, distinct inputs always produce distinct outputs.

- $f$ is surjective if its range is the whole codomain, and hence given any $y \in B$ there is some $x \in A$ such that $f(x) = y$.

**Example 1.4.** The function $f : \mathbb{R} \to \mathbb{R}$ which maps $x$ to $f(x) = x^3$ is injective:

$$f(x_1) = f(x_2) \qquad \Rightarrow x_1^3 = x_2^3$$
$$\Rightarrow x_1^3 - x_2^3 = 0$$
$$\Rightarrow (x_1 - x_2)(x_1^2 + x_1 x_2 + x_2^2) = 0$$

which implies that $x_1 = x_2$, or $x_1^2 + x_1 x_2 + x_2^2 = 0$. But

$$x_1^2 + x_1 x_2 + x_2^2 = (x_1 + \frac{1}{2}x_2)^2 + \frac{3}{4}x_2^2 .$$

which is the sum of two non-negative numbers and, hence, is zero if and only if $x_2 = 0$ and $x_1 + \frac{1}{2}x_2 = 0$. But then $x_1 = x_2 = 0$. So, in every case, $f(x_1) = f(x_2)$ implies that $x_1 = x_2$. Hence $f$ is injective.

In fact, $f$ is also surjective, though we are not (yet) in a position to prove this. What this means is that, given any real number $y$, there is some real number $x$ such that $x^3 = y$. You may say "this is obvious: $x = y^{\frac{1}{3}}$", but what is $y^{\frac{1}{3}}$? It is, by definition, the real number whose cube is $y$. How do you know that such a real number exists? You don't (yet). □

**Definition 1.5.** Given functions $f : A \to B$ and $g : B \to C$, their **composition** is the function $g \circ f : A \to C$ which assigns to each $x \in A$ the image $(g \circ f)(x) = g(f(x))$.

So $g \circ f$ is the function obtained by first "doing $f$" then taking the output and feeding it into $g$.[1] Note that we need the codomain of $f$ to match the domain of $g$ in order for $g(f(x))$ to make sense, otherwise $f(x)$ might not be in the domain of $g$, in which case $g(f(x))$ would be undefined.

**Example 1.6.** Let $f : \mathbb{R} \to \mathbb{R}$ such that $f(x) = x^2$ and $g : \mathbb{R} \to \mathbb{R}$ such that $g(x) = x + 1$. Then $g \circ f : \mathbb{R} \to \mathbb{R}$ such that

$$(g \circ f)(x) = g(f(x)) = g(x^2) = x^2 + 1.$$

In this case, the composition in the opposite order (first $g$ then $f$) is also well-defined, $f \circ g : \mathbb{R} \to \mathbb{R}$,

$$(f \circ g)(x) = f(g(x)) = f(x + 1) = (x + 1)^2.$$

Clearly $g \circ f \neq f \circ g$. □

---

[1] The fact that function composition is ordered from right to left is a legacy of algebra's roots in the Arabic-speaking world.

Function composition preserves both injectivity and surjectivity, as we now prove.

**Proposition 1.7.** *Let $f : A \to B$ and $g : B \to C$.*

*(i) If $f$ and $g$ are both injective, so is $g \circ f$.*
*(ii) If $f$ and $g$ are both surjective, so is $g \circ f$.*

**Proof.** (i) Assume $f$ and $g$ are injective, and that $(g \circ f)(x_1) = (g \circ f)(x_2)$ for some $x_1, x_2 \in A$. Then $g(f(x_1)) = g(f(x_2))$. But $g$ is injective, so it follows that $f(x_1) = f(x_2)$. But $f$ is injective, so it follows that $x_1 = x_2$. Hence $g \circ f$ is injective.

(ii) Assume $f$ and $g$ are surjective, and let $z \in C$. Since $g$ is surjective, $z$ is in the range of $g$, so there exists $y \in B$ such that $g(y) = z$. But $f$ is also surjective, so $y$ is in the range of $f$, and hence there exists $x \in A$ such that $f(x) = y$. Now $(g \circ f)(x) = g(f(x)) = g(y) = z$. Hence $z$ is in the range of $g \circ f$. This argument works for any $z \in C$, so $g \circ f$ is surjective.

$\square$

**Definition 1.8.** Given any set $A$, the **identity map** (or **identity function**) on $A$ is the function

$$\mathrm{Id}_A : A \to A, \qquad \mathrm{Id}_A(x) = x.$$

This is the function that maps every element of $A$ to itself.

**Definition 1.9.** A function $g : B \to A$ is an **inverse function** for $f : A \to B$ if $g \circ f = \mathrm{Id}_A$ and $f \circ g = \mathrm{Id}_B$. In this case we say that $f$ is **invertible**.

Note that, if $g$ is an inverse function for $f$, then $f$ is an inverse function for $g$.

**Example 1.10.** Let $f : \mathbb{R} \to \mathbb{R}$, $f(x) = 2x - 1$. Then $g : \mathbb{R} \to \mathbb{R}$, $g(x) = (x+1)/2$ is an inverse function to $f$ (and *vice versa*):

$$(g \circ f)(x) = g(2x - 1) = ((2x - 1) + 1)/2 = x$$
$$(f \circ g)(x) = f((x+1)/2) = 2\left(\frac{x+1}{2}\right) - 1 = x.$$

$\square$

In general, it is not enough to check just one of the conditions

$$g \circ f = \mathrm{Id}_A, \qquad f \circ g = \mathrm{Id}_B,$$

as the next example illustrates. To state it, we need the *absolute value* function on $\mathbb{R}$.

**Definition 1.11.** The **absolute value** of a real number $x$ is

$$|x| = \begin{cases} x \text{ if } x \geq 0, \\ -x \text{ if } x < 0. \end{cases}$$

By definition, $|x| \geq 0$ for all $x$.

**Example 1.12.** Consider the pair of functions $f : \mathbb{R} \to [0, \infty)$, $f(x) = |x|$ and $g : [0, \infty) \to \mathbb{R}$, $g(x) = x$. Then, for all $x \in [0, \infty)$,

$$(f \circ g)(x) = f(x) = |x| = x$$

since $x \geq 0$. Hence, $f \circ g = \mathrm{Id}_{[0,\infty)}$. However,

$$(g \circ f)(-1) = g(|-1|) = g(1) = 1 \neq -1$$

so $g \circ f \neq \mathrm{Id}_{\mathbb{R}}$. Hence, $g$ is not an inverse function for $f$, nor is $f$ an inverse function for $g$. □

**Proposition 1.13.** *If $f : A \to B$ is invertible, then its inverse function is unique.*

**Proof.** Let $g : B \to A$ and $g' : B \to A$ both be inverses for $f$. Then, for all $y \in B$,

$$f(g(y)) = \ y \ = f(g'(y))$$
$$\Rightarrow \quad g(f(g(y))) = g(y) = g(f(g'(y)))$$
$$\Rightarrow \quad g(y) = g(y) = g'(y)$$

since $g \circ f = \mathrm{Id}_A$. Since $g'(y) = g(y)$ for all $y \in B$, the functions $g$ and $g'$ coincide. □

So it makes sense to speak of *the* inverse of an invertible function. It is common to denote the inverse of $f$ by $f^{-1}$. Be careful not to confuse $f^{-1}$ with $1/f$. For example, $f : \mathbb{R} \to \mathbb{R}$, $f(x) = x$ is invertible, and its inverse is $f^{-1}(x) = x = f(x)$, which has nothing to do with $x^{-1} = 1/x$. How can we tell whether a given function is invertible?

**Proposition 1.14.** $f : A \to B$ *is invertible if and only if it is bijective.*

**Proof.** We first prove the "if" direction. So assume that $f$ is bijective. Then, for each $y \in B$, there exists (since $f$ is surjective) a unique (since $f$ is injective) $x \in A$ such that $f(x) = y$. Denote by $g : B \to A$ the mapping which maps each $y$ to its corresponding $x$. Then, for all $y \in B$, $f(g(y)) = y$, and for all $x \in A$, $g(f(x)) = x$, by the definition of $g$. Hence, $f \circ g = \mathrm{Id}_B$ and $g \circ f = \mathrm{Id}_A$.

We now prove the "only if" direction. So assume that a function $g : B \to A$ exists such that $g \circ f = \mathrm{Id}_A$ and $f \circ g = \mathrm{Id}_B$. Assume $f(x) = f(x')$ for some $x, x' \in A$. Then $g(f(x)) = g(f(x'))$. But $g \circ f = \mathrm{Id}_A$, so $x = x'$. Hence, $f$ is injective. Finally, let $y \in B$. Then $g(y) \in A$ and $f(g(y)) = y$ since $f \circ g = \mathrm{Id}_B$. Hence, $f$ is surjective. $\qquad\square$

Returning to Example 1.12, we see that $f$ is not bijective, since it is not injective (for example, $f(-1) = f(1)$). Hence, by Proposition 1.14, it is not invertible, so $g$ cannot be its inverse.

## 1.3 Boundedness and the Axiom of Completeness

When we are first taught about real numbers, it is usual to think of them as infinite decimal expansions, that is, expressions like

$$261070.040172230103141005\ldots$$

with finitely many digits taken from $\{0, 1, 2, \ldots, 9\}$ to the left of a point and infinitely many digits to the right of it (possibly all but finitely many of these being 0). This is OK, although it has some disadvantages (the choice of base 10 for the expansion is pretty arbitrary, and there's the unfortunate fact that different expansions can represent the same number, e.g. $0.9999\ldots = 1.0000\ldots$). We will take a very different approach, much better suited to our purposes.

Rather than *define* the set of real numbers $\mathbb{R}$, we will, for the time being,[2] characterize it by stating what properties we assume it has. First, we assume that $\mathbb{R}$ is a **field**, that is, a set on which the four basic operations of arithmetic $(+, \times, -, \div)$ are well-defined and have the familiar properties (for example $x + y = y + x$, and $x \times (y + z) = (x \times y) + (x \times z)$). Second, we assume that $\mathbb{R}$ is, in fact, an **ordered** field: it is equipped with an ordering relation $<$, also satisfying standard properties (for example, if $x < y$ then $x + z < y + z$ for all $z$). These structures (the arithmetic operations and the

---

[2]A constructive definition of the reals is given in Chapter 13, but this is rather advanced material and is not required to understand the rest of the book.

ordering relation) have exactly the properties you've been familar with since elementary school, so we will not labour them here, although the interested reader will find the precise definition of an ordered field in Chapter 13 (Definitions 13.10 and 13.12).

Another example of an ordered field is $\mathbb{Q}$, the set of rational numbers. So what is the difference between $\mathbb{Q}$ and $\mathbb{R}$? For our purposes, the key difference is that $\mathbb{R}$ satisfies an extra property called **completeness**, while $\mathbb{Q}$ does not. This property is not familar from school level mathematics, so we will study it in some depth. To define completeness requires some preliminary ideas.

**Definition 1.15.** Let $A$ be a subset of $\mathbb{R}$. An **upper bound** on $A$ is any $K \in \mathbb{R}$ such that, for all $x \in A$, $x \leq K$. A **lower bound** on $A$ is any $L \in \mathbb{R}$ such that, for all $x \in A$, $x \geq L$. If $A$ has an upper bound, we say that $A$ is **bounded above**. If $A$ has a lower bound, we say that $A$ is **bounded below**. $A$ is **bounded** if it is bounded both above and below.

**Example 1.16.** The set $A = [-1, 2)$ is bounded above, by 2 for example, and below, by $-1$ for example. Hence, $A$ is bounded. Note that upper and lower bounds are *not* unique. That is, $2, 43, 19345\pi$ are all upper bounds on $A$. In fact, any $K \geq 2$ is an upper bound on $A$, and any $L \leq -1$ is a lower bound on $A$. □

It is important not to confuse *bounded* with *finite*. A finite set is one containing a finite number of distinct elements. Any finite subset of $\mathbb{R}$ will certainly be bounded, but the converse is clearly false: not every bounded subset of $\mathbb{R}$ is finite, as the above example shows.

**Example 1.17.** The set of positive integers $\mathbb{Z}^+$ is bounded below, by 1 for example (or $-43$, or $-1000$). However, it is *not* bounded above. □

Let's think carefully about what it means to say that a set $A$ is *unbounded above*. It means that $A$ has no upper bound. Equivalently, every real number $K$ is *not* an upper bound on $A$. Hence, given any $K \in \mathbb{R}$, there exists $x \in A$ such that $x > K$.

So an alternative formulation of the statement "$\mathbb{Z}^+$ is unbounded above" is "given any real number $K$ there exists a positive integer $n$ such that $n > K$". We will use this simple observation over and over again, so it's worth giving it a name:

**Fact 1.18 (The Archimedean Property of $\mathbb{R}$).** Given   any   real number $K$, there is some positive integer $n$ such that $n > K$.

Note that $\mathbb{Q}$ also has the Archimedean Property (given any $K \in \mathbb{Q}$ there is some $n \in \mathbb{Z}^+$ such that $n > K$).

**Definition 1.19.** The **supremum** of a subset $A$ of $\mathbb{R}$ is its least upper bound (if this exists), denoted $\sup A$. So $\sup A$ is a real number with two properties:

(i) $\sup A$ is an upper bound on $A$.

(ii) No number less than $\sup A$ is an upper bound on $A$.

Similarly the **infimum** of $A$ is its greatest lower bound (if this exists), denoted $\inf A$.

It follows immediately from the definition that $\sup A$, if it exists, is unique. For if two numbers both satisfy the above two properties and are different, then the smaller of them, $K_1$ say, is an upper bound on $A$ by (i), and is less than the larger one, $K_2$ say, contradicting (ii). So it makes sense to speak of *the* supremum and/or infimum of a subset of $\mathbb{R}$, in contrast to upper/lower bounds.

**Example 1.20.** Let $A = [-1, 2)$. Then, immediately from the definition of this kind of interval, $\sup A = 2$ and $\inf A = -1$.     □

**Example 1.21.** Let $A = \{\frac{1}{n} : n \in \mathbb{Z}^+\}$. Every element of $A$ is positive, so $A$ is bounded below, by 0. In fact, $\inf A = 0$. Let's prove this. We already know that 0 is a lower bound on $A$, so it just remains to show that every number bigger than 0 is *not* a lower bound on $A$. So, let $K > 0$. Then $1/K$ is a real number, so by the Archimedean Property, there exists some $n \in \mathbb{Z}^+$ such that $n > 1/K$. But then $1/n \in A$ and $1/n < K$, so $K$ is not a lower bound on $A$.

The supremum is a bit less subtle. I claim that $\sup A = 1$. Again, let's prove this. Let $x \in A$. Then $x = 1/n$ for some $n \in \mathbb{Z}^+$. But $n \geq 1$, so $x \leq 1$. Hence 1 is an upper bound on $A$. Any $K < 1$ is not an upper bound on $A$, since $1 \in A$.     □

The existence of suprema (= plural of supremum) is the defining difference between $\mathbb{R}$ and $\mathbb{Q}$.

**Axiom 1.22 (The Axiom of Completeness).** Every nonempty subset of $\mathbb{R}$ which is bounded above has a supremum in $\mathbb{R}$.

The equivalent statement for $\mathbb{Q}$ (replace $\mathbb{R}$ by $\mathbb{Q}$ in both places) is false. One could show this directly, by considering the set

$$A = \{x \in \mathbb{Q} : x^2 < 2\}$$

for example. This is a nonempty subset of $\mathbb{Q}$ (e.g. $1 \in A$) which is bounded above (e.g. every $x \in A$ is less than 2), but it has no supremum in $\mathbb{Q}$, that is, there is no *rational* number $r$ with the required properties that (i) $r$ is an upper bound on $A$ and (ii) no rational number $s < r$ is an upper bound on $A$. To show this, one would prove that every rational upper bound $r$ on $A$ has $r^2 > 2$ and that, given any such bound, a smaller rational number (for example $s = rn/(n+1)$, where $n$ is a sufficiently large positive integer) exists which is also an upper bound on $A$. Constructing a direct proof along these lines is an instructive exercise which the interested reader is invited to attempt (the argument should, at some point, appeal to Proposition 1.1, of course). We will give an indirect proof of incompleteness of $\mathbb{Q}$ shortly.

The Axiom of Completeness looks rather lopsided: it says that every nonempty subset of $\mathbb{R}$ which is bounded *above*, has a supremum, but says nothing about sets which are bounded *below*. That's because the equivalent statement for *infima* ($=$ plural of infimum) follows directly from Axiom 1.22. Proving this is a good exercise in keeping clear sight of what upper/lower bound and sup/inf mean:

**Proposition 1.23.** *Let $A \subset \mathbb{R}$ be nonempty and bounded below. Then $A$ has an infimum.*

**Proof.** Let $B = \{-x : x \in A\}$, which is certainly nonempty. Let $K$ be a lower bound on $A$. Then $-K$ is an upper bound on $B$. (Check: if $y \in B$ then $-y \in A$, and hence $-y \geq K$. Hence $y \leq -K$.) Hence, $B$ is a nonempty subset of $\mathbb{R}$ which is bounded above so, by the Axiom of Completeness, $B$ has a supremum, $L$ say. Now $-L$ is a lower bound on $A$. (Check: if $x \in A$ then $-x \in B$, so $-x \leq L$. Hence $x \geq -L$.) Let $M$ be any real number greater than $-L$. Then $-M < L$, and $L$ is the *least* upper bound on $B$. Hence $-M$ is *not* an upper bound on $B$, so there exists $y \in B$ such that $y > -M$. But then $-y \in A$ and $-y < M$, so $M$ is not a lower bound on $A$. Hence, no number greater than $-L$ is a lower bound on $A$. Hence $-L$ is the infimum of $A$. $\qquad\square$

The following lemma will be used later, in Chapter 11. Its proof is also a good exercise in applying the definitions of supremum and infimum.

**Lemma 1.24.** *Let $A \subset \mathbb{R}$ be bounded (above and below) and define* $\operatorname{diff} A$ *to be the set*

$$\operatorname{diff} A = \{x - y \ : \ x, y \in A\}.$$

*Then* $\operatorname{diff} A$ *is bounded above and*

$$\sup \operatorname{diff} A = \sup A - \inf A.$$

**Proof.** Let $z \in \operatorname{diff} A$. Then there exist $x, y \in A$ such that $x - y = z$. Now $x \leq \sup A$ and $y \geq \inf A$, so $-y \leq -\inf A$, whence

$$x - y \leq \sup A + (-\inf A).$$

This is true for all $z \in \operatorname{diff} A$, so $\operatorname{diff} A$ is bounded above by $\sup A - \inf A$.

Now, let $K$ be any real number less than $\sup A - \inf A$. Let

$$\varepsilon = K - (\sup A - \inf A) > 0.$$

Then $\sup A - \varepsilon/2 < \sup A$, so is *not* an upper bound on $A$. Hence, there exists $x \in A$ such that $x > \sup A - \varepsilon/2$. Similarly, $\inf A + \varepsilon/2 > \inf A$, so is *not* a lower bound on $A$. Hence, there exists $y \in A$ such that $y < \inf A + \varepsilon/2$. But then $x - y \in \operatorname{diff} A$, and

$$x - y > (\sup A - \frac{\varepsilon}{2}) - (\inf A + \frac{\varepsilon}{2}) = K.$$

So $K$ is not an upper bound on $\operatorname{diff} A$. This is true for all $K < \sup A - \inf A$, and $\sup A - \inf A$ *is* an upper bound on $\operatorname{diff} A$, so we conclude that $\sup \operatorname{diff} A = \sup A - \inf A$. $\qquad\square$

## 1.4 Some consequences of the Axiom of Completeness

In this section, we will deduce some basic properties of the set $\mathbb{R}$ by using Axiom 1.22. The first of these we have already asserted as Fact 1.18: the set $\mathbb{Z}^+$ is unbounded above. This looks entirely obvious, but is it really? As far as we're concerned, $\mathbb{R}$ is, *by definition*, a complete ordered field (so we are not defining it to be the set of all decimal expansions, for example). Is it really obvious that a complete ordered field cannot have a magic element, $x_\infty$ say, which is bigger than every positive integer?[3]

---

[3]The sceptical reader might ask "how do I know that this complete ordered field contains the set of positive integers at all?" Well, to be a field, it must contain a distinguished element $1_{\mathbb{R}}$ whose defining property is that $1_{\mathbb{R}} \times x = x$ for all $x \in \mathbb{R}$. We can identify this special element with $1 \in \mathbb{Z}^+$. But then, since $\mathbb{R}$ is closed under addition, it contains $1 + 1$, $1 + 1 + 1$, etc., and, by the properties of the ordering relation, these are all distinct elements. Arguing similarly, one may show that $\mathbb{R}$ must contain $\mathbb{Z}$ and $\mathbb{Q}$.

**Theorem 1.18 (The Archimedean Property of $\mathbb{R}$).** *Given any real number $K$, there is some positive integer $n$ such that $n > K$.*

**Proof.** Assume, to the contrary, that $\mathbb{Z}^+$ is bounded above. Then, by Axiom 1.22, $\mathbb{Z}^+$ has a supremum, $U$ say. Since $U - 1 < U$, it is not an upper bound on $\mathbb{Z}^+$, so there exists $n \in \mathbb{Z}^+$ such that $n > U - 1$. But then $n + 1 \in \mathbb{Z}^+$, and $n + 1 > U$, contradicting the fact that $U$ is an upper bound on $\mathbb{Z}^+$. $\qquad\square$

That's a relief! As a corollary, we can prove that the set of rational numbers must be **dense** in the set of reals, in the following sense:

**Theorem 1.25 ($\mathbb{Q}$ is dense in $\mathbb{R}$).** *Between any pair of distinct real numbers, there is a rational number.*

The proof will require the following lemma, whose proof is almost identical to the proof of Theorem 1.18. Note that the only property of $\mathbb{Z}$ we used in this proof was closure under addition of 1, that is, if $m \in \mathbb{Z}^+$ then $m + 1 \in \mathbb{Z}^+$.

**Lemma 1.26.** *Let $E \subseteq \mathbb{R}$ be nonempty and have the property that whenever $m \in E$ then $m - 1 \in E$. Then $E$ is unbounded below.*

**Proof.** Let $E$ have the property stated and assume, towards a contradiction that $E$ is bounded below. Then, by Proposition 1.23, $E$ has an infimum, $K$ say. Since $K + 1 > K$ it is, by the definition of infimum, not a lower bound on $E$. Hence, there exists $m \in E$ such that $m < K + 1$. But then $m - 1 \in E$ also, and $m - 1 < K$, contradicting the definition of $K$. $\qquad\square$

*Proof of Theorem 1.25:* Let's call the smaller of the two real numbers $x$ and the larger $y$. By the Archimedean Property of $\mathbb{R}$ (Theorem 1.18), there exists $n \in \mathbb{Z}^+$ such that $n > \frac{1}{y-x}$, and hence, $ny - 1 > nx$. Consider the set

$$E = \{m \in \mathbb{Z} : m > nx\}.$$

This is nonempty, by the Archimedean Property of $\mathbb{R}$ again (if it weren't, $nx$ would be an upper bound on $\mathbb{Z}$, and hence on $\mathbb{Z}^+ \subset \mathbb{Z}$). We claim there exists $m \in E$ such that $m < ny$. Assume this is false, so every $m \in E$ satisfies $m \geq ny$. Then, given any $m \in E$, $m - 1 \geq ny - 1 > nx$, so $m - 1 \in E$. Hence, by Lemma 1.26, $E$ is unbounded below. But $E$ is clearly bounded below (by $nx$) by definition, a contradiction. Hence, the claim is true: there exists $m \in \mathbb{Z}$ with $nx < m < ny$. But then $x < \frac{m}{n} < y$, as was to be proved. □

It's not hard (and is a good exercise) to show that between any pair of distinct real numbers there are *infinitely many* rational numbers. In fact, how do we know there are any elements of $\mathbb{R}$ that are *not* rational numbers? One way to settle the issue is to prove that $\mathbb{Q}$ is *countable*, while $\mathbb{R}$ is not.

**Definition 1.27.** A set $A$ is **countable** if there exists a surjective function $f : \mathbb{Z}^+ \to A$.

Given such a function $f : \mathbb{Z}^+ \to A$, we can imagine writing down an infinite list

$$f(1)$$
$$f(2)$$
$$f(3)$$
$$\vdots$$

of elements of $A$. Since $f$ is surjective, this list contains each and every element of $A$ (possibly more than once).

**Example 1.28.**

(i) Clearly $\mathbb{Z}^+$ is countable: we just take $f : \mathbb{Z}^+ \to \mathbb{Z}^+$, $f(n) = n$.

(ii) Less obviously, $\mathbb{Z}$ is also countable. This is slightly surprising at first sight, because $\mathbb{Z}$ naively looks "twice as big" as $\mathbb{Z}^+$ (it contains $\mathbb{Z}^+$, but much more, namely $0$ and the set of negative integers). However, it is easy to construct a surjective function $f : \mathbb{Z}^+ \to \mathbb{Z}$. For example

$$f(n) = \begin{cases} \frac{n}{2} - 1 & \text{if } n \text{ is even}, \\ -\frac{n+1}{2} & \text{if } n \text{ is odd}, \end{cases}$$

will do. This function maps the odd positive integers to $-1, -2, -3, \ldots$ in turn, and the even positive integers to $0, 1, 2, \ldots$ in turn. □

Naively, $\mathbb{Q}$ looks much bigger than $\mathbb{Z}^+$. Nonetheless, it is countable.

**Theorem 1.29.** *The set $\mathbb{Q}$ of rational numbers is countable.*

**Proof.** We first show that the set of *positive* rational numbers

$$\mathbb{Q}_+ = \{\frac{n}{m} : n, m \in \mathbb{Z}^+\}$$

is countable. To do this, we imagine constructing a two dimensional table, extending infinitely downwards and to the right. The rows of this table are labelled by a positive integer $n$ and the columns are labelled by a positive integer $m$. In the box at position $(n, m)$, that is, on row $n$ and column $m$, we enter the rational number $\frac{n}{m}$. It is clear that every positive rational number appears somewhere in this table (multiple times in fact: for example boxes $(2, 3)$, $(3, 9)$ and $(4, 12)$ all contain the rational number $\frac{2}{3}$). We now construct a surjective function $g$ from $\mathbb{Z}^+$ to the set of boxes in the table, as illustrated in Figure 1.1. The map that assigns to each positive integer

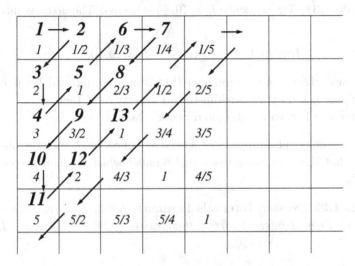

Fig. 1.1   Constructing a surjective map $\mathbb{Z}^+ \to \mathbb{Q}_+$.

$k$ the rational number in box $g(k)$ is a surjective map $\mathbb{Z}^+ \to \mathbb{Q}_+$. Hence, there exists a surjective map $f : \mathbb{Z}^+ \to \mathbb{Q}_+$, that is, $\mathbb{Q}_+$ is countable.

Consider now the function $h : \mathbb{Z} \to \mathbb{Q}$ defined by

$$h(n) = \begin{cases} f(n) & \text{if } n > 0, \\ 0 & \text{if } n = 0, \\ -f(-n) & \text{if } n < 0, \end{cases}$$

where $f : \mathbb{Z}^+ \to \mathbb{Q}_+$ is the surjective function just constructed. The function $h$ is also surjective (check it!). Furthermore, $\mathbb{Z}$ is countable (see Example 1.28), so there exists a surjective function $F : \mathbb{Z}^+ \to \mathbb{Z}$. But then $h \circ F : \mathbb{Z}^+ \to \mathbb{Q}$ is surjective (Proposition 1.7), so $\mathbb{Q}$ is countable. $\qquad \square$

We next prove that the set $\mathbb{R}$ is *uncountable*. It's not hard to show that the set of all decimal expansions is uncountable. But, as far as we're concerned, $\mathbb{R}$ is just a complete ordered field and there is no obvious reason why every element of such a field should be representable by a decimal expansion. To prove that $\mathbb{R}$ is uncountable without the crutch of decimal expansions, we need to introduce the idea of nested intervals.

**Definition 1.30.** A sequence of intervals $I_1, I_2, I_3, \ldots$ is said to be **nested** if $I_{n+1} \subseteq I_n$ for all $n \in \mathbb{Z}^+$.

**Example 1.31.** The sequence $I_n = (0, \frac{1}{n}]$ is nested. The first few intervals are

$$I_1 = (0, 1], \quad I_2 = (0, \frac{1}{2}], \quad I_3 = (0, \frac{1}{3}], \ldots.$$

In this case, there is no real number that is in every interval $I_n$: if $x$ is in $I_1$ then $x > 0$, and by the Archimedean Property of $\mathbb{R}$, there exists $k \in \mathbb{Z}^+$ such that $k > 1/x$, so $x > 1/k$, and hence $x$ fails to be in $I_k$. $\qquad\square$

It is a fundamental property of nested sequences of *closed* intervals $I_n = [a_n, b_n]$ that there always exists a real number which is simultaneously in every $I_n$.

**Lemma 1.32 (Nested Intervals Lemma).** *Let $I_n = [a_n, b_n]$ be a nested sequence of closed intervals. Then there exists $x \in \mathbb{R}$ such that $x \in I_n$ for all $n \in \mathbb{Z}^+$.*

**Proof.** Consider the set $A = \{a_n : n \in \mathbb{Z}^+\}$, which is certainly nonempty. Let $m \in \mathbb{Z}^+$ and consider $b_m$, the upper endpoint of $I_m$. Let $n \in \mathbb{Z}^+$. If $n \geq m$ then, by the nesting property, $[a_n, b_n] \subseteq [a_m, b_m]$, so $a_n \in [a_m, b_m]$ and hence $a_n \leq b_m$. If $n < m$ then, by the nesting property, $[a_m, b_m] \subseteq [a_n, b_n]$, so $b_m \in [a_n, b_n]$ and hence $b_m \geq a_n$. Hence, for all $n \in \mathbb{Z}^+$, $a_n \leq b_m$, that is, $b_m$ is an upper bound on $A$. Since $A$ is nonempty and bounded above, it has, by the Axiom of Completeness, a supremum, $x$ say. We claim that $x \in I_n$ for all $n$. Since $x$ is an upper bound on $A$, $x \geq a_n$ for all $n \in \mathbb{Z}^+$. Further, as we just showed, every $b_n$ is an upper bound on $A$, so $x \leq b_n$ for all $n \in \mathbb{Z}^+$ (else it isn't the *least* upper bound on $A$). Hence $x \in [a_n, b_n] = I_n$ for all $n \in \mathbb{Z}^+$. $\qquad\square$

It's not hard to deduce from this lemma that $\mathbb{R}$ must be uncountable. In fact, we'll show that $[0, 1]$ is uncountable, from which the uncountability of $\mathbb{R}$ immediately follows.

**Theorem 1.33.** $[0,1]$ *is uncountable.*

**Proof.** Assume, to the contrary, that there is a surjective function $f : \mathbb{Z}^+ \to [0,1]$. Let $I_1$ be any closed interval in $[0,1]$ which does not contain $f(1)$, $I_2$ be any closed subinterval of $I_1$ which does not contain $f(2)$, and so on. In this way, we obtain a nested sequence of closed intervals $I_n$ such that, for each $n \in \mathbb{Z}^+$, $f(n) \notin I_n$. By Lemma 1.32, there exists $x \in [0,1]$ such that $x \in I_n$ for all $n \in \mathbb{Z}^+$. Hence, $x \neq f(n)$ for all $n \in \mathbb{Z}^+$, that is, $x$ is not in the range of $f$. But $f$ is surjective, a contradiction. $\square$

The only extra property we assumed $\mathbb{R}$ possesses beyond those which $\mathbb{Q}$ certainly also possesses (an ordering relation and the basic arithmetic operations) is the Axiom of Completeness, and this one extra property was enough to conclude that $\mathbb{R}$ is uncountable. Since $\mathbb{Q}$ is countable, it must follow that it does *not* satisfy the Axiom of Completeness. So we can take Theorems 1.29 and 1.33 as an indirect proof that $\mathbb{Q}$ is incomplete.

We say that a real number is **irrational** if it is not rational. Since $\mathbb{Q}$ is countable, and the union of any pair of countable sets is countable (exercise: prove it!), we deduce immediately that the set of irrational numbers, $\mathbb{R}\backslash\mathbb{Q}$, is uncountable. In particular, there certainly *are* irrational numbers in $[0,1]$, although we do not yet have a single explicit example of such a number! (We proved in Proposition 1.1 that $\sqrt{2}$ can't be rational, but we have *not* yet proved that it's irrational because we haven't yet proved it exists at all! That is, we have not yet proved that there exists a real number whose square is 2. We'll return to this later.) From the existence of a single irrational number, we can easily deduce that the irrational numbers are, like the rational numbers, dense in $\mathbb{R}$:

**Theorem 1.34 ($\mathbb{R}\backslash\mathbb{Q}$ is dense in $\mathbb{R}$).** *Between any pair of distinct real numbers there is an irrational number.*

**Proof.** Let's call the smaller of the two real numbers $x$ and the larger $y$. Let $w$ be an irrational number in $[0,1]$. Then $w > 0$, so by Theorem 1.25, there exists a rational number $\frac{m}{n}$ such that

$$\frac{x}{w} < \frac{m}{n} < \frac{y}{w}.$$

Hence $\frac{m}{n}w$ lies between $x$ and $y$ and is certainly irrational (since if $\frac{m}{n}w = \frac{p}{q}$ then $w = \frac{pn}{qm}$, which is rational). $\square$

So, between any pair of rationals there's an irrational, and between any pair of irrationals there's a rational. But there are more irrationals between 0 and 1 than there are rationals in $\mathbb{R}$! This set $\mathbb{R}$ is very, very strange...

## 1.5 Summary

- An **upper bound** on a set $A \subset \mathbb{R}$ is a number $K \in \mathbb{R}$ such that $x \leq K$ for all $x \in A$. A **lower bound** on $A$ is $L \in \mathbb{R}$ such that $x \geq L$ for all $x \in A$. $A$ is **bounded above** if it has an upper bound, **bounded below** if it has a lower bound, and **bounded** if it has both.
- The **supremum** of $A$ is its least upper bound. That is, $K = \sup A$ if (i) $K$ is an upper bound on $A$ and (ii) no $L < K$ is an upper bound on $A$.
- The **infimum** of $A$ is its greatest lower bound. That is, $K = \inf A$ if (i) $K$ is a lower bound on $A$ and (ii) no $L > K$ is a lower bound on $A$.
- $\mathbb{R}$ is, by definition, an ordered field which additionally satisfies the **Axiom of Completeness:** every nonempty subset of $\mathbb{R}$ which is bounded above has a supremum in $\mathbb{R}$.
- It follows from this definition that:

  - $\mathbb{R}$ has the Archimedean Property: given any $K \in \mathbb{R}$, there exists $n \in \mathbb{Z}^+$ such that $n > K$.
  - $\mathbb{Q}$ is dense in $\mathbb{R}$: between any pair of distinct real numbers, there is a rational number.
  - $\mathbb{R} \backslash \mathbb{Q}$ is dense in $\mathbb{R}$: between any pair of distinct real numbers, there is an irrational number.
  - $\mathbb{R} \backslash \mathbb{Q}$ is uncountable.

## 1.6   Tutorial problems

1. Rewrite the set $A = \mathbb{R}\backslash(([-3,2] \cup [4,7))\backslash[0,5))$ as a union of intervals.
2. (a) Prove that $f : \mathbb{R}\backslash\{-1\} \to \mathbb{R}\backslash\{0\}$, $f(x) = \dfrac{1}{x+1}$ is bijective.
   (b) Prove that $g : \mathbb{R} \to \mathbb{R}$, $g(x) = x^3 - x$ is not injective.
3. Let $f : A \to B$ and $g : B \to A$. Prove that:

   (a) If $g \circ f = \mathrm{Id}_A$ and $f$ is surjective then $f \circ g = \mathrm{Id}_B$.
   (b) If $f \circ g = \mathrm{Id}_B$ and $f$ is injective then $g \circ f = \mathrm{Id}_A$.
4. Determine whether the set $A = \{x^2 + 2x + y \ : \ x \in \mathbb{R}, y \in [-1,1]\}$ is bounded above, bounded below, both, or neither. Rigorously justify your answer.
5. Let $A = \{\dfrac{1}{n} - \dfrac{2}{m} \ : \ n \in \mathbb{Z}^+, m \in \mathbb{Z}^+\}$.

   (a) Prove that $\sup A = 1$.
   (b) Is $1 \in A$?
   (c) Does $A$ have an infimum? If so, what is it?

## 1.7   Homework problems

1. Rewrite each of the following subsets of $\mathbb{R}$ as an interval or union of intervals:

$$A = \{x \in \mathbb{R} \ : \ x^2 + 2x \neq 0\},$$
$$B = \mathbb{R}\backslash([-1,1) \cup (2,3)),$$
$$C = [-43,43]\backslash((-100,10) \cap [-10,100]),$$
$$D = \{x + y \ : \ x \in \mathbb{R}, y \in \mathbb{R}, -1 \leq x \leq 1, y > 0\}.$$

2. (a) Prove that $f : \mathbb{R} \to \mathbb{R}$, $f(x) = (x + \dfrac{1}{3})^2$ is not injective.

   (b) Prove that $g : \mathbb{Z} \to \mathbb{Q}$, $g(x) = (x + \dfrac{1}{3})^2$ is injective.
3. Let $f : A \to B$ and $g : B \to A$.

   (a) Assume that $g \circ f = \mathrm{Id}_A$ and $f$ is injective. Does it follow that $f \circ g = \mathrm{Id}_B$?
   (b) Assume that $f \circ g = \mathrm{Id}_B$ and $f$ is surjective. Does it follow that $g \circ f = \mathrm{Id}_A$?
4. Determine whether the following subsets of $\mathbb{R}$ are bounded above, bounded below, both, or neither. Rigorously justify your answers.

(a) $A = \{x + \dfrac{1}{x} : x \in (0, \infty)\}$

(b) $B = \{x^2 + xy^2 : -2 \le x \le 1, -1 \le y \le 1\}$.

5. Let $A = \{\dfrac{n}{m} : n \in \mathbb{Z}^+, m \in \mathbb{Z}^+, m > n\}$.

(a) Prove that $\inf A = 0$.

(b) Prove that $\sup A = 1$.

# Chapter 2

# Real sequences

## 2.1 Definition and examples of real sequences

A **real sequence** is a mapping $a : \mathbb{Z}^+ \to \mathbb{R}$, that is, a rule which assigns to each positive integer $n$ some real number $a(n)$. Actually, for sequences, unlike pretty much all other functions, it's conventional to write the image of $n$ under the mapping $a$ as $a_n$ instead of $a(n)$, and call it the $\boldsymbol{n^{\text{th}}}$ **term** of the sequence, rather than the "image of $n$". It's also conventional to denote the mapping as $(a_n)_{n \in \mathbb{Z}^+}$, or just $(a_n)$ for short, rather than $a :$ $\mathbb{Z}^+ \to \mathbb{R}$. We will follow these conventions. Sometimes, it's convenient to call mappings $a : \mathbb{N} \to \mathbb{R}$ sequences too, i.e. to start numbering the terms at $n = 0$ instead of $n = 1$.

Often we will specify a sequence just by giving a formula for its $n^{\text{th}}$ term.

**Example 2.1.** $a_n = \frac{n^2 + 5}{n^2}$ is a sequence whose first four terms are

$$6, \quad \frac{9}{4}, \quad \frac{14}{9}, \quad \frac{21}{16}.$$

As $n$ gets very large, $a_n$ gets very close to 1: for example $a_n = 1$ to three decimal places for all $n \geq 101$. So for $n$ this large, we can approximate the sequence by the constant number 1, provided we're happy to accept errors no larger than $5 \times 10^{-4}$. What if we're a bit fussier than this and need to know $a_n$ to at least five decimal places? How big must $n$ be before we can approximate $a_n$ by the constant 1? We need

$$0.999995 \le a_n < 1.000005$$

$$\Leftrightarrow \quad 0.999995 \le 1 + \frac{5}{n^2} < 1.000005$$

$$\Leftrightarrow \quad -0.000005 \le \frac{5}{n^2} < 0.000005$$

$$\Leftrightarrow \quad \frac{5}{n^2} < 5 \times 10^{-6}$$

$$\Leftrightarrow \quad n^2 > 10^6$$

$$\Leftrightarrow \quad n > 1000.$$

So, for all $n \ge 1001$, $a_n = 1$ to five decimal places. In fact, however many decimal places of accuracy we insist on, there is a point in the sequence beyond which, to that accuracy, $a_n$ is just 1. □

**Example 2.2.** Contrast this with $a_n = \sin n$, whose first four terms (to five decimal places) are $0.84147, 0.90930, 0.14112$ and $-0.75680$. The behaviour of this sequence as $n$ grows large is not at all uniform; the terms bounce around seemingly at random in the interval $[-1, 1]$. □

Sometimes it's more convenient to define a sequence **inductively**. In its simplest form, this is when we specify the first term $a_1$ and give a rule for constructing $a_{n+1}$ from $a_n$.

**Example 2.3.** Let $(a_n)$ be the sequence whose first term is $a_1 = 1$, for which

$$a_{n+1} = \frac{a_n}{1 + a_n^2}.$$

The first four terms of this sequence are

$$a_1 = 1,$$

$$a_2 = \frac{1}{1 + 1^2} = \frac{1}{2},$$

$$a_3 = \frac{1/2}{1 + (1/4)} = \frac{2}{5},$$

$$a_4 = \frac{2/5}{1 + (4/25)} = \frac{10}{29}.$$

What happens to $a_n$ as $n$ gets very large? This is actually quite hard to answer directly, even using a computer, because the terms become very unwieldy as $n$ gets even moderately large. For example

$$a_8 = \frac{2684403867986594189884909}{10993317375225483680390211},$$

and my computer has trouble computing even $a_{30}$ because the integers involved get so large. We can convert each $a_n$ to a truncated decimal expansion, which greatly speeds up the calculation, but this gives rounding errors which accumulate as $n$ increases, so by the time we get to $a_{1000}$, say, it's not obvious whether the answer my computer produces (0.02234293705, for what it's worth) bears any relation to reality.

Another interesting question is whether the large $n$ behaviour of this sequence depends on our choice of initial term, $a_1 = 1$. What if $a_1 = 0$? Or $a_1 = 1000$? We will develop methods which will allow us to show that, whatever $a_1$ we choose, $a_n$ for large $n$ becomes very close to 0 – despite the fact that we have no idea how to write down $a_n$ in general! □

In Example 2.1, the terms of the sequence "tend to" 1 as $n$ gets arbitrarily large, whereas in Example 2.2, the terms bounce around indefinitely, without tending to a particular value. We say that $a_n = (n^2 + 5)/n^2$ *converges* to 1, while $a_n = \sin n$ does not converge. It is now time to make this concept of convergence precise.

## 2.2   Convergence of a real sequence

Recall (Definition 1.11) that the *absolute value* $|x|$ of a real number $x$ is $x$ if $x \geq 0$ and $-x$ if $x < 0$. So $|4| = 4$, $|-2| = -(-2) = 2$ and so on. Clearly $|x| \geq 0$ for all $x$, by definition. Furthermore, by checking the various cases ($x$ negative, $x$ non-negative etc.) it's easy to show that

$$|x| \geq x \tag{2.1}$$
$$|xy| = |x||y|. \tag{2.2}$$

If we think of real numbers as lying on an infinite straight line, then $|x|$ can be interpreted geometrically as the *distance* from $x$ to 0. Furthermore, given any *pair* of real numbers $x, y$, the absolute value of their difference $|x - y|$ can be thought of as the distance between them. Note that $|x-y| = |y-x|$ for all $x$ and $y$.

For example, the distance between 4 and $-2$ is

$$|4 - (-2)| = |6| = 6 \quad \text{or} \quad |(-2) - 4| = |-6| = -(-6) = 6.$$

We emphasize that this is just a mental model of $\mathbb{R}$ which gives a nice interpretation of $|\cdot|$. We are not *defining* real numbers as points on an infinite line, nor are we *defining* $|x|$ to be the distance from $x$ to 0.

The following useful property of absolute values follows quickly from the equations (2.1), (2.2):

**Proposition 2.4 (The Triangle Inequality).** *For all $x, y \in \mathbb{R}$, $|x+y| \leq |x| + |y|$.*

**Proof.** Assume, to the contrary, that $|x + y| > |x| + |y|$. Since both sides are non-negative, it follows that

$$|x + y|^2 > |x + y|(|x| + |y|) > (|x| + |y|)^2$$
$$\Rightarrow \quad (x + y)^2 > |x|^2 + 2|x||y| + |y|^2$$
$$\Rightarrow \quad x^2 + 2xy + y^2 > x^2 + 2|xy| + y^2 \qquad \text{by equation (2.2)}$$
$$\Rightarrow \quad xy > |xy|$$

which contradicts inequality (2.1). $\qquad\qquad\qquad\qquad\qquad\qquad\qquad\square$

What does this have to do with convergence of sequences? In saying that $a_n = \frac{(n^2+5)}{n^2}$ converges to 1, we are saying, roughly, that the terms $a_n$ are close to 1 for large $n$. In terms of absolute values, this means that $|a_n - 1|$ is small for large $n$. More precisely, we make the following definition:

**Definition 2.5.** A real sequence $(a_n)$ **converges** to a real number $L$ if, for each $\varepsilon > 0$, there exists some positive integer $N$ such that, for all $n \geq N$, $|a_n - L| < \varepsilon$. In this case, we will write $a_n \to L$. The number $L$ is called the **limit** of the sequence $(a_n)$ and is often denoted $\lim a_n$ or $\lim_{n\to\infty} a_n$. If $(a_n)$ does not converge (to any limit), we say it **diverges**.

### Remarks

- Definition 2.5 is the most important definition in this book, and it is important to understand it thoroughly. It is the first rigorous definition of limit we have encountered. Other definitions of limit (e.g. $\lim_{x\to a} f(x)$) will be constructed from it.
- The definition says that given *any* $\varepsilon > 0$, no matter how small, there is a term in the sequence, $a_N$, after which *all* terms in the sequence lie within distance $\varepsilon$ of $L$.
- Note that $|a_n - L| < \varepsilon$ is equivalent to $L - \varepsilon < a_n < L + \varepsilon$, or, $a_n \in (L - \varepsilon, L + \varepsilon)$.
- If we imagine plotting the terms of the sequence $a_n$ on a graph (with $n$ along the $x$-axis and $a_n$ along the $y$-axis), we can construct a geometric picture of what the definition means. Given any positive number $\varepsilon$, construct the horizontal strip of height $2\varepsilon$ centred on the line $y = L$. Then there exists a point $N$ on the $x$-axis, to the right of which all points on the graph lie in the strip.

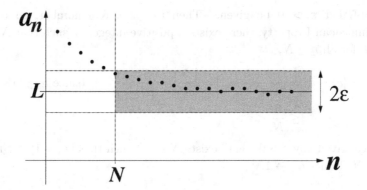

This is true no matter how small we make $\varepsilon$. If we make $\varepsilon$ smaller, the strip gets narrower, and the $N$ beyond which all terms lie in the strip (probably) moves rightwards.

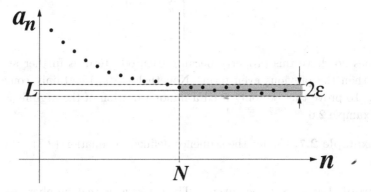

- So $N$ will, in general, depend on $\varepsilon$: if $\varepsilon$ is large, a fairly small $N$ will probably work, whereas if $\varepsilon$ is very small, $N$ will probably have to be very large. For our purposes, the important thing is to establish that, given any $\varepsilon > 0$, there is a corresponding $N$ which works for that particular $\varepsilon$. The relationship between $\varepsilon$ and $N$ is not important, nor is it important to know the smallest $N$ which will work for a given $\varepsilon$.

A *proof from first principles* that a given sequence $(a_n)$ converges to a given limit $L$ is a direct argument showing that, given any $\varepsilon > 0$, there is some $N \in \mathbb{Z}^+$ such that whenever $n \geq N$, $|a_n - L| < \varepsilon$. The key to doing this is to "estimate" (that is, rigorously find an upper bound on) the quantity $|a_n - L|$.

**Example 2.6.** Claim: $a_n = \frac{n^2+5}{n^2} \to 1$.

**Proof.** Let $\varepsilon > 0$ be given. Then $5/\varepsilon$ is a real number, so by the Archimedean Property, there exists a positive integer $N$ such that $N > \frac{5}{\varepsilon}$. Now, for all $n \geq N$,

$$|a_n - 1| = \left| \frac{n^2 + 5}{n^2} - 1 \right| = \left| \frac{5}{n^2} \right| = \frac{5}{n^2} \leq \frac{5}{n} \qquad \text{(since } n \geq 1)$$

$$\leq \frac{5}{N} < \varepsilon.$$

Hence, given any $\varepsilon > 0$, there exists $N \in \mathbb{Z}^+$ such that $|a_n - 1| < \varepsilon$ for all $n \geq N$. Hence $a_n \to 1$. $\qquad \square$

Note that Definition 2.5 does *not* say that $a_n$ "gets closer and closer to $L$ as $n$ gets larger", because this isn't necessarily true. For example, the sequence

$$a_n = \begin{cases} \frac{1}{n} & n \text{ odd} \\[2mm] \frac{1}{n^2} & n \text{ even} \end{cases} \qquad (2.3)$$

does not have this property because each odd term is further away from 0 than the previous even term. Nonetheless, $(a_n)$ certainly converges to 0. To prove this, we only need a minor variation of the argument used in Example 2.6.

**Example 2.7.** Claim: the sequence defined in equation (2.3) converges to 0.

**Proof.** Let $\varepsilon > 0$ be given. Then $1/\varepsilon$ is a real number, so by the Archimedean Property, there exists a positive integer $N$ such that $N > \frac{1}{\varepsilon}$. Now, for all $n \geq N$,

$$|a_n - 0| = a_n \qquad \text{(since } a_n > 0 \text{ for all } n)$$

$$= \begin{cases} \frac{1}{n} & n \text{ odd} \\[2mm] \frac{1}{n^2} & n \text{ even} \end{cases}$$

$$\leq \frac{1}{n} \qquad \text{(since } n \geq 1)$$

$$\leq \frac{1}{N}$$

$$< \varepsilon.$$

Hence, given any $\varepsilon > 0$, there exists $N \in \mathbb{Z}^+$ such that $|a_n - 0| < \varepsilon$ for all $n \geq N$. Hence $a_n \to 0$. $\qquad \square$

Note that Definition 2.5 does *not* merely say that, given any $\varepsilon > 0$, there is some $N \in \mathbb{Z}^+$ such that $|a_N - L| < \varepsilon$. The definition certainly implies this, but is considerably stronger. For example, consider the sequence $a_n = (-1)^n$. Then, given any $\varepsilon > 0$, there is an $N \in \mathbb{Z}^+$ such that $|a_n - 1| < \varepsilon$. For example, $N = 2$ will do for any $\varepsilon > 0$, since $|a_2 - 1| = |1 - 1| = 0 < \varepsilon$. But clearly $a_n$ does not converge to 1. In fact, $a_n$ does not converge to anything. Can we prove this? Yes:

**Example 2.8.** Claim: the sequence $a_n = (-1)^n$ does not converge (to *any* limit).

**Proof.** Assume, to the contrary, that $a_n \to L$. Then, given any positive number $\varepsilon$, there exists $N \in \mathbb{Z}^+$ such that $|a_n - L| < \varepsilon$ for all $n \geq N$. In particular, this must be true in the case $\varepsilon = 1$: there exists $N \in \mathbb{Z}^+$ such that, for all $n \geq N$, $|a_n - L| < 1$. But then $N$ and $N + 1$ are consecutive integers, so one is odd and the other is even, and hence $|a_{N+1} - a_N| = 2$. But, by the Triangle Inequality,

$$|a_{N+1} - a_N| = |(a_{N+1} - L) - (a_N - L)| \leq |a_{N+1} - L| + |a_N - L| < 1 + 1 = 2$$

by the definition of $N$. Hence $2 < 2$, a contradiction. $\square$

We next establish that two rather simple, but important, sequences converge to their expected limits.

**Example 2.9.** The constant sequence, $a_n = c$ for all $n$, converges to $c$.

**Proof.** Given any $\varepsilon > 0$ we can take $N = 1$, since for all $n \geq 1$, $|a_n - c| = |c - c| = 0 < \varepsilon$. $\square$

**Example 2.10.** The sequence $a_n = \frac{1}{n}$ converges to 0.

**Proof.** Given any $\varepsilon > 0$, there exists $N > \frac{1}{\varepsilon}$ by the Archimedean Property. Then, for all $n \geq N$, $|a_n - 0| = \frac{1}{n} \leq \frac{1}{N} < \varepsilon$. $\square$

Direct proofs of this kind are often called "$\varepsilon$–$N$ proofs". We finish this section by showing how a direct $\varepsilon$–$N$ proof can be made, even for a reasonably complicated sequence.

**Example 2.11.** Claim: $a_n = \dfrac{n^2 - \sin n}{(2n - 7)(3n + 1)} \to \dfrac{1}{6}$.

**Proof.** The key is to estimate $|a_n - \frac{1}{6}|$. For all $n \in \mathbb{Z}^+$,

$$\left|a_n - \frac{1}{6}\right| = \frac{1}{6}\left|\frac{6n^2 - 6\sin n - (2n-7)(3n+1)}{(2n-7)(3n+1)}\right|$$

$$= \frac{1}{6}\frac{|6n^2 - 6\sin n - (6n^2 - 19n - 7)|}{|2n-7||3n+1|}$$

$$= \frac{1}{6}\frac{|19n - 6\sin n + 7|}{(2n-7)(3n+1)} \qquad \text{(if } n \geq 4)$$

$$\leq \frac{1}{6}\frac{19n + 6|\sin n| + 7}{(2n-7)(3n+1)} \qquad \text{(Triangle Inequality)}$$

$$\leq \frac{1}{6}\frac{32n}{(2n-7)(3n+1)} \qquad (|\sin n| \leq 1 \text{ and } 1 \leq n)$$

$$< \frac{1}{6}\frac{32n}{(2n-n)3n} \qquad \text{(if } n \geq 7)$$

$$< \frac{2}{n}$$

So we've shown that for all $n \geq 7$, $|a_n - \frac{1}{6}| < \frac{2}{n}$.

Now, let $\varepsilon > 0$ be given. Then, by the Archimedean Property of $\mathbb{R}$, there exists $N \in \mathbb{Z}^+$ such that $N \geq \max\{\frac{2}{\varepsilon}, 7\}$. Then for all $n \geq N$, $n \geq 7$ and $n \geq \frac{2}{\varepsilon}$, so by the above estimate

$$\left|a_n - \frac{1}{6}\right| < \frac{2}{n} \leq \varepsilon.$$

$\square$

## 2.3 Summary

- A **real sequence** is a mapping $a : \mathbb{Z}^+ \to \mathbb{R}$. We denote the image of $n \in \mathbb{Z}^+$ under the mapping $a$ by $a_n$ and call it the $n^{th}$ **term** of the sequence. We usually denote the sequence as a whole by $(a_n)$.
- A sequence $(a_n)$ **converges** to a real number $L$ if, for each $\varepsilon > 0$, there exists a positive integer $N$ such that, for all $n \geq N$, $|a_n - L| < \varepsilon$.
- If $(a_n)$ converges to $L$, we write $a_n \to L$.
- The number $L$ is called the **limit** of the sequence, sometimes denoted $\lim a_n$ or $\lim_{n \to \infty} a_n$.
- An $\varepsilon$–$N$ proof of convergence is a direct argument showing that, given any positive number $\varepsilon$, there is a positive integer $N$ such that $|a_n - L| < \varepsilon$ for all $n \geq N$.

## 2.4   Tutorial problems

1. Prove from first principles that the following sequences converge. (Hint: your first job is to figure out, informally, what the limit is in each case.)

$$a_n = \frac{2n^2 - n}{n^2 + 2}, \quad b_n = \frac{1 - (-1)^n + 6n^2}{7 - 2n^2}, \quad c_n = \sqrt{n+1} - \sqrt{n}, \quad d_n = \frac{2^n}{n!}$$

For the purposes of $(c_n)$ you can suspend disbelief and assume that square roots exist (recall we haven't actually proved this yet!).

2. Let $\beta$ be a positive constant. Prove by induction that $(1+\beta)^n \geq 1 + n\beta$, for all $n \in \mathbb{Z}^+$. Hence prove (from first principles) that the sequence

$$a_n = \frac{1}{(1 + \beta)^n} \quad \text{converges to 0.}$$

## 2.5   Homework problems

1. Let $a_n = \dfrac{5n}{4n - 3}$. This sequence converges to $L = \frac{5}{4}$.

   (a) Find the smallest $N \in \mathbb{Z}^+$ such that $|a_n - L| < 10^{-4}$ for all $n \geq N$.
   (b) Show that, for all $n > 12$, $|a_n - L| < 1/n$. Deduce an integer $M$ (*not necessarily the smallest!*) such that $|a_n - L| < 10^{-6}$ for all $n \geq M$.
   (c) Prove from first principles that $a_n$ converges to $\frac{5}{4}$.

2. Prove from first principles that the following sequences converge. (Hint: your first job is to figure out, informally, what the limit is in each case.)

$$a_n = \frac{n+1}{n+2}, \qquad\qquad b_n = \frac{5n}{4n^2 - 3},$$

$$c_n = \begin{cases} 400/n & n \text{ odd} \\ -1/(400n^2) & n \text{ even} \end{cases}, \qquad d_n = \begin{cases} n/(1+n^2) & 1 \leq n \leq 900 \\ 7 & n \geq 901 \end{cases}.$$

# Chapter 3

# Limit theorems

## 3.1 Some basic limit theorems

It's very important to get plenty of practice in proving directly from first principles that sequences converge (i.e. constructing $\varepsilon$–$N$ proofs) because this is the best way to get to grips with what Definition 2.5 really means. It is also good preparation for understanding the proofs of the following basic limit theorems. Once we've proved these, we can often use them, together with a few standard examples such as Examples 2.9 and 2.10, to show indirectly that a given sequence converges, without having to give an $\varepsilon$–$N$ argument.

**Proposition 3.1 (Uniqueness of Limits).** *If a sequence converges, its limit is unique.*

**Proof.** Assume, to the contrary, that a sequence $a_n$ converges to both $L_1$ and $L_2$ with $L_1 \neq L_2$. Let $\varepsilon = \frac{1}{2}|L_1 - L_2| > 0$. Since $a_n \to L_1$, there exists $N_1 \in \mathbb{Z}^+$ such that, for all $n \geq N_1$,
$$|a_n - L_1| < \varepsilon.$$
Similarly, since $a_n \to L_2$, there exists $N_2 \in \mathbb{Z}^+$ such that, for all $n \geq N_2$,
$$|a_n - L_2| < \varepsilon.$$
Now let $n$ be any positive integer greater than both $N_1$ and $N_2$ (for example $n = N_1 + N_2$). Then both the above inequalities hold, and so,
$$2\varepsilon = |L_2 - L_1| = |(a_n - L_1) - (a_n - L_2)|$$
$$\leq |a_n - L_1| + |a_n - L_2| \quad \text{(by the Triangle Inequality)}$$
$$< \varepsilon + \varepsilon \quad \text{(since } n > N_1 \text{ and } n > N_2\text{),}$$
that is, $2\varepsilon < 2\varepsilon$, a contradiction. Hence, the original assumption (that $(a_n)$ has two different limits) must be false. $\qquad\square$

So we are justified in speaking of *the* limit of a convergent sequence.

**Definition 3.2.** We say that a real sequence $(a_n)$ is **bounded above** if its range, the set $A = \{a_n : n \in \mathbb{Z}^+\}$ is bounded above (as in Definition 1.15), that is, if there exists $K \in \mathbb{R}$ such that $a_n \leq K$ for all $n \in \mathbb{Z}^+$. Similarly, $(a_n)$ is **bounded below** if $A$ is bounded below, and $(a_n)$ is **bounded** if it is bounded both above and below.

**Proposition 3.3.** *If a sequence converges then it is bounded.*

**Proof.** Assume $a_n \to L$. Then, given any $\varepsilon > 0$, there exists $N \in \mathbb{Z}^+$ such that, for all $n \geq N$, $|a_n - L| < \varepsilon$. This is true, in particular, in the case $\varepsilon = 1$. Hence, there exists $N \in \mathbb{Z}^+$ such that, for all $n \geq N$, $|a_n - L| < 1$. Let

$$K = \max\{|a_1 - L|, |a_2 - L|, \ldots, |a_{N-1} - L|, 1\},$$

which exists since the set is finite. Then for all $n \in \mathbb{Z}^+$, $|a_n - L| \leq K$, and hence, $L - K \leq a_n \leq L + K$. Hence, the sequence $(a_n)$ is bounded (above by $L + K$ and below by $L - K$). $\qquad\square$

Of course, the converse of this Proposition is false: not every bounded sequence is convergent, as we have already seen (for example $a_n = (-1)^n$ is bounded, but not convergent). *If* a sequence is convergent, it's bounded above and below, and it's not hard to show that its limit must lie between the upper and lower bounds:

**Proposition 3.4.** *If $K \leq a_n \leq M$ for all $n \in \mathbb{Z}^+$ and $(a_n)$ converges to $L$, then $K \leq L \leq M$.*

**Proof.** Exercise. The obvious strategy is a proof by contradiction. That is, assume $a_n \to L$ but $L \notin [K, M]$. Then either (i) $L > M$ or (ii) $L < K$. Now, in each of these cases, show that a contradiction arises. $\qquad\square$

Note that a sequence $(a_n)$ is bounded if and only if $|a_n|$ is bounded above.

**Theorem 3.5 (Algebra of Limits).** *If $a_n \to A$ and $b_n \to B$ then*

*(i)* $a_n + b_n \to A + B$,
*(ii)* $a_n b_n \to AB$.

*If, in addition, $a_n \neq 0$ for all $n$ and $A \neq 0$, then*

(iii) $\dfrac{1}{a_n} \to \dfrac{1}{A}$.

**Proof.** (i) Let $\varepsilon > 0$ be given. Then $\frac{\varepsilon}{2}$ is also a positive number, and $a_n \to A$, so there exists $N_1 \in \mathbb{Z}^+$ such that, for all $n \geq N_1$, $|a_n - A| < \frac{\varepsilon}{2}$. Similarly, $b_n \to B$, so there exists $N_2 \in \mathbb{Z}^+$ such that, for all $n \geq N_2$, $|b_n - B| < \frac{\varepsilon}{2}$. Let $N = \max\{N_1, N_2\}$. Then, for all $n \geq N$,

$$
\begin{aligned}
|(a_n + b_n) - (A + B)| &= |(a_n - A) + (b_n - B)| \\
&\leq |a_n - A| + |b_n - B| \quad \text{(Triangle Inequality)} \\
&< \frac{\varepsilon}{2} + \frac{\varepsilon}{2} \quad \text{(since } n \geq N_1 \text{ and } n \geq N_2) \\
&= \varepsilon.
\end{aligned}
$$

Hence, given any $\varepsilon > 0$, there exists $N \in \mathbb{Z}^+$ such that, for all $n \geq N$, $|a_n + b_n - (A + B)| < \varepsilon$. Hence $a_n + b_n \to A + B$.

(ii) Let $\varepsilon > 0$ be given. By Proposition 3.3, since $b_n$ is convergent, it is bounded, so there exists $K > 0$ such that $|b_n| \leq K$ for all $n$. Then $\varepsilon' = \varepsilon/(K + |A|)$ is a positive number, and $a_n \to A$ and $b_n \to B$, so there exist $N_1, N_2 \in \mathbb{Z}^+$ such that $|a_n - A| < \varepsilon'$ for all $n \geq N_1$, and $|b_n - B| < \varepsilon'$ for all $n \geq N_2$. Let $N = \max\{N_1, N_2\}$. Then for all $n \geq N$,

$$
\begin{aligned}
|a_n b_n - AB| &= |a_n b_n - Ab_n + Ab_n - AB| \\
&= |(a_n - A)b_n + A(b_n - B)| \\
&\leq |a_n - A||b_n| + |A||b_n - B| \quad \text{(Triangle Inequality)} \\
&\leq |a_n - A|K + |A||b_n - B| \\
&< \varepsilon' K + |A|\varepsilon' \quad \text{(since } n \geq N_1 \text{ and } n \geq N_2) \\
&= \varepsilon.
\end{aligned}
$$

Hence, $a_n b_n \to AB$.

(iii) Let $\varepsilon > 0$ be given. Since $a_n \to A \neq 0$, and $\frac{|A|}{2}$ is a positive number, there exists $N_1 \in \mathbb{Z}^+$ such that, for all $n \geq N_1$, $|a_n - A| < \frac{|A|}{2}$, and hence, $|a_n| > \frac{|A|}{2}$. Similarly, $\varepsilon' = \frac{1}{2}|A|^2$ is also a positive number, so there exists $N_2 \in \mathbb{Z}^+$, such that, for all $n \geq N_2$, $|a_n - A| < \varepsilon'$. Let

$N = \max\{N_1, N_2\}$. Then for all $n \geq N$,

$$\left| \frac{1}{a_n} - \frac{1}{A} \right| = \left| \frac{A - a_n}{a_n A} \right| = \frac{|a_n - A|}{|a_n||A|}$$

$$< \frac{\varepsilon'}{|a_n||A|} \qquad (\text{since } n \geq N_2)$$

$$< \frac{\varepsilon'}{(|A|/2)|A|} \qquad (\text{since } n \geq N_1)$$

$$= \varepsilon.$$

Hence, $\frac{1}{a_n} \to \frac{1}{A}$.

$\square$

Starting from simple known facts, like $1/n \to 0$, we can use Theorem 3.5 to deduce the convergence of a large collection of sequences. For example, $1/n^2$, $1/n^3$, $1/n^4 \ldots$ all converge to 0:

**Example 3.6.** Claim: for any integer $p \geq 1$, the sequence $a_n = \dfrac{1}{n^p}$ converges to 0.

**Proof.** We will prove this by induction (on $p$). The claim is true for $p = 1$, by Example 2.10. Assume the claim holds for $p = k$. Then, in the case $p = k + 1$,

$$a_n = \frac{1}{n}\frac{1}{n^k} \to 0 \times 0 = 0$$

by Theorem 3.5 (part (ii)). Hence, by induction, the claim holds for all $p \in \mathbb{Z}^+$. $\square$

**Example 3.7.** Claim: the sequence $a_n = \dfrac{2n^2 + n}{n^2 + 7}$ converges to 2.

**Proof.** First note that

$$a_n = \frac{2 + \frac{1}{n}}{1 + \frac{7}{n^2}} = \left( 2 + \frac{1}{n} \right) \times \frac{1}{1 + \frac{7}{n^2}}.$$

Now the constant sequence $(2, 2, 2, \ldots)$ converges to 2 (Example 2.9) and $\frac{1}{n} \to 0$ (Example 2.10), so $2 + \frac{1}{n} \to 2 + 0 = 2$ by Theorem 3.5(i). Similarly, 1 converges to 1, 7 to 7 and $\frac{1}{n^2}$ converges to 0 (Example 3.6), so $7/n^2 \to 7 \times 0 = 0$ (Theorem 3.5(ii)), and hence $1 + \frac{7}{n^2} \to 1$ (Theorem 3.5(i)). Furthermore, $1 + \frac{7}{n^2}$ is never 0, so $1/(1 + \frac{7}{n^2}) \to \frac{1}{1} = 1$ (Theorem 3.5(iii)). Hence $a_n \to 2 \times 1 = 2$ (Theorem 3.5(ii)). $\square$

Clearly, if we maintain this level of detail in our arguments, we will pass out through exhaustion. So, in future, we will use Theorem 3.5 more freely, along the lines of the following example:

**Example 3.8.** Claim: the sequence $a_n = \dfrac{n^4}{2(n+1)^2(n^2+1)}$ converges to $\frac{1}{2}$.

**Proof.**

$$a_n = \frac{1}{2(1 + (1/n))^2(1 + (1/n^2))} \to \frac{1}{2 \times (1+0)^2 \times (1+0)} = \frac{1}{2}$$

by Theorem 3.5. $\qquad\square$

Exercise 3.9. Prove from first principles that the sequence in Example 3.8 converges to $\frac{1}{2}$ (i.e. give a direct $\varepsilon$–$N$ proof). It's good for your (mathematical) soul...

The Algebra of Limits (Theorem 3.5) allows us to prove convergence of many sequences by splitting them up into simpler pieces whose limits we already know. Sometimes it's easier to use less detailed, more generic properties of a sequence. Our final limit theorem in this section is an extremely useful and versatile result: the Squeeze Rule. This says that if a sequence is trapped ("squeezed") between two other sequences which are known to converge to a common limit, then the trapped sequence also converges (to the same limit).

**Theorem 3.10 (The Squeeze Rule).** *Let $(a_n)$, $(b_n)$ and $(c_n)$ be sequences such that $(a_n)$ and $(c_n)$ converge to the same limit $L$. If $a_n \le b_n \le c_n$ for all $n$, then $(b_n)$ converges to $L$ also.*

**Proof.** Let $\varepsilon > 0$ be given. Since $a_n \to L$, there exists $N_1 \in \mathbb{Z}^+$ such that, for all $n \ge N_1$, $|a_n - L| < \varepsilon$, and hence, $a_n > L - \varepsilon$. Similarly, since $c_n \to L$, there exists $N_2 \in \mathbb{Z}^+$ such that, for all $n \ge N_2$, $|c_n - L| < \varepsilon$, and hence, $c_n < L + \varepsilon$. Let $N = \max\{N_1, N_2\}$. Then for all $n \ge N$,

$$b_n - L \ge a_n - L$$
$$> -\varepsilon \quad \text{(since } n \ge N_1)$$

and $\quad b_n - L \le c_n - L$
$$< \varepsilon \quad \text{(since } n \ge N_2).$$

Hence, for all $n \ge N$, $|b_n - L| < \varepsilon$. $\qquad\square$

**Example 3.11.** Claim: $b_n = \dfrac{\cos n}{n^2 + 1} \to 0$.

**Proof.** Let $a_n = -\dfrac{1}{n^2}$ and $c_n = \dfrac{1}{n^2}$. Then $a_n \leq b_n \leq c_n$ for all $n \in \mathbb{Z}^+$ (since $-1 \leq \cos n \leq 1$) and $a_n \to 0$ and $c_n \to 0$ (by Example 3.6 and Theorem 3.5), so $b_n \to 0$ by the Squeeze Rule. $\qquad\square$

## 3.2 The Monotone Convergence Theorem

None of the limit theorems we have proved so far have made any use of the Axiom of Completeness. We next prove a very fundamental result which does: the Monotone Convergence Theorem. Like the Squeeze Rule, this is a theorem which allows us to prove convergence of some sequences using only rough, generic information about them.

**Definition 3.12.** A sequence $(a_n)$ is **increasing** if $a_{n+1} \geq a_n$ for all $n \in \mathbb{Z}^+$. It is **decreasing** if $a_{n+1} \leq a_n$ for all $n \in \mathbb{Z}^+$. It is **monotone** if it is increasing or decreasing.

**Remarks**

- It follows immediately from the definition that if $(a_n)$ is increasing then $a_n \geq a_m$ whenever $n \geq m$. Similarly, a decreasing sequence has $a_n \leq a_m$ for all $n \geq m$.
- It follows that an increasing sequence is automatically bounded below (by $a_1$) and a decreasing sequence is automatically bounded above (again, by $a_1$).
- Note that it's possible for a sequence to be both increasing and decreasing! But only if it's constant.
- We sometimes refer to a sequence as **strictly increasing**, meaning $a_{n+1} > a_n$ for all $n \in \mathbb{Z}^+$.

We have proved that all convergent sequences are bounded and observed that not all bounded sequences are convergent ($a_n = (-1)^n$ provides a counterexample). We next prove that all *monotone* bounded sequences *are* convergent.

**Theorem 3.13.** *Let $(a_n)$ be increasing and bounded above. Then $(a_n)$ converges.*

**Proof.** The set $A = \{a_n : n \in \mathbb{Z}^+\}$ is nonempty and bounded above, and hence, by the Axiom of Completeness, has a supremum, $L \in \mathbb{R}$. We claim that $a_n \to L$.

Let $\varepsilon > 0$ be given. Then $L - \varepsilon < L$, so by the definition of supremum, $L - \varepsilon$ is *not* an upper bound on $A$. Hence there exists $N \in \mathbb{Z}^+$ such that $a_N > L - \varepsilon$. But $(a_n)$ is increasing, so for all $n \geq N$, $a_n \geq a_N > L - \varepsilon$. Further, $L$ *is* an upper bound on $A$, so for all $n \in \mathbb{Z}^+$, $a_n < L < L + \varepsilon$. Putting these two inequalities together, we see that for all $n \geq N$

$$-\varepsilon < a_n - L < \varepsilon$$

and hence $|a_n - L| < \varepsilon$. $\qquad\square$

Note that, as already remarked, the proof of this theorem relies on the Axiom of Completeness. That every decreasing sequence bounded below converges follows immediately from Theorem 3.13 and the Algebra of Limits.

**Corollary 3.14 (Monotone Convergence Theorem).** *Every bounded monotone sequence converges.*

**Proof.** Let $(a_n)$ be bounded and monotone. If $(a_n)$ is increasing, it converges by Theorem 3.13, so we may assume that $(a_n)$ is decreasing. Let $b_n = -a_n$. Then $(b_n)$ is increasing and bounded, so converges by Theorem 3.13. Hence $(a_n) = (-b_n)$ converges by the Algebra of Limits. $\qquad\square$

**Example 3.15.** Let $a_n = \alpha^n$ where $\alpha \in (0, 1)$. Claim: $a_n \to 0$.

**Proof.** $a_n > 0$ for all $n$, and $a_{n+1} = \alpha a_n < a_n$ since $0 < \alpha < 1$, so $(a_n)$ is decreasing and bounded below. Hence, by the Monotone Convergence Theorem, $a_n \to L$ for some $L$. But $b_n = a_{n+1}$ converges to $L$ also (it's the same sequence, but with the first term omitted), and $b_n = \alpha a_n$, so by Theorem 3.5, $b_n \to \alpha L$. But the limit of a convergent sequence is unique (Theorem 3.1), so $L = \alpha L$. If $L \neq 0$, then $\alpha = L/L = 1$, a contradiction. Hence $L = 0$. $\qquad\square$

The key trick in the proof above was to think of $(a_n)$ as an inductively defined sequence: $a_1 = \alpha$, $a_{n+1} = \alpha a_n$. We can try the same strategy for other inductively defined sequences.

**Example 2.3 (revisited).** Recall we defined a sequence $(a_n)$ inductively by specifying $a_1$ and using the rule

$$a_{n+1} = \frac{a_n}{1 + a_n^2} \tag{3.1}$$

for all $n \in \mathbb{Z}^+$. Consider this sequence in the case where $a_1 = \alpha > 0$ (the case $\alpha < 0$ can be handled similarly, and the case $\alpha = 0$ is trivial). We claim that $a_n \to 0$.

**Proof.** First note that $a_n > 0$ for all $n$, as one can easily show by induction ($a_1 > 0$ and if $a_n > 0$ then $a_{n+1} > 0$ by equation (3.1)). It follows, since

$$\frac{a_{n+1}}{a_n} = \frac{1}{1 + a_n^2} < 1,$$

that $a_{n+1} < a_n$ for all $n$. Hence $(a_n)$ is decreasing. We have already shown that $(a_n)$ is bounded below (by 0), so by the Monotone Convergence Theorem, $a_n$ converges to some limit $L$. Clearly, the sequence $b_n = a_{n+1}$ also converges to $L$ (it's the same sequence but with the first term omitted). But

$$b_n = \frac{a_n}{1 + a_n^2}$$

so, by the Algebra of Limits, $b_n$ converges to $L/(1 + L^2)$. But the limit of a convergent sequence is unique (Theorem 3.1), so

$$L = \frac{L}{1 + L^2}$$

whose only solution is $L = 0$. Hence $a_n \to 0$. $\qquad\square$

## 3.3 Sequences and suprema

A recurrent theme in this book is that we formulate all the fundamental notions of real analysis in terms of sequences and their convergence properties. As an early example of this, we can give a sequential characterization of suprema and infima.

**Proposition 3.16.** *Let $A$ be a nonempty subset of $\mathbb{R}$ and $K \in \mathbb{R}$. Then*

*(i) $K = \sup A$ if and only if $K$ is an upper bound on $A$ and there exists a sequence $(x_n)$ in $A$ such that $x_n \to K$.*

*(ii) $K = \inf A$ if and only if $K$ is a lower bound on $A$ and there exists a sequence $(x_n)$ in $A$ such that $x_n \to K$.*

**Proof.** We will prove part (i) and leave part (ii) as an exercise.

Let $K = \sup A$. Then $K$ is an upper bound on $A$. For each $n \in \mathbb{Z}^+$, $K - 1/n < K$ so, by the definition of supremum, cannot be an upper bound on $A$. Hence, for each $n \in \mathbb{Z}^+$, there exists $x_n \in A$ such that $x_n > K - \frac{1}{n}$. Since $x_n \in A$, $x_n \leq K$ ($K$ *is* an upper bound on $A$). Hence, for all $n \in \mathbb{Z}^+$, $K - \frac{1}{n} < x_n \leq K$. Now $K - \frac{1}{n} \to K$ and the constant sequence $K \to K$, so, by the Squeeze Rule, $x_n \to K$ also.

Conversely, assume that $K$ is an upper bound on $A$ and there exists a sequence $(x_n)$ in $A$ such that $x_n \to K$. Let $L < K$. Then $\varepsilon = K - L > 0$, and $x_n \to K$, so there exists $N \in \mathbb{Z}^+$ such that, for all $n \geq N$, $|x_n - K| < \varepsilon$. In particular, $x_N - K > -\varepsilon$, so $x_N > K - \varepsilon = L$. Hence, there exists an element of $A$, namely $x_N$, which is greater than $L$, so $L$ is not an upper bound on $A$. Since $K$ is an upper bound on $A$, and no number smaller than $K$ is an upper bound on $A$, we conclude that $K = \sup A$. □

**Example 3.17.** Let $A = \{\frac{2m}{n^2} : n, m \in \mathbb{Z}^+, n < m\}$. I claim that $\inf A = 0$.

**Proof.** Every element of $A$ is positive, so 0 is a lower bound on $A$. Now, for each $n \in \mathbb{Z}^+$,

$$x_n = \frac{2(n+1)}{n^2} \in A$$

(just take $m = n + 1$). By the Algebra of Limits,

$$x_n = \frac{2}{n} + \frac{2}{n^2} \to 0 + 0 = 0$$

so $\inf A = 0$ by Proposition 3.16. □

## 3.4 Summary

- The limit of a convergent sequence is unique.
- If a sequence is convergent, then it is bounded. The converse is false.
- The usual **Algebra of Limits** holds: if $a_n \to A$ and $b_n \to B$ then
    - $a_n + b_n \to A + B$,
    - $a_n b_n \to AB$, and
    - $a_n / b_n \to A/B$ (provided $B \neq 0$ and $b_n$ is never 0).
- The **Squeeze Rule**: if $a_n \leq b_n \leq c_n$ and $a_n \to L$ and $c_n \to L$, then $b_n \to L$.
- A sequence $(a_n)$ is **increasing** if $a_{n+1} \geq a_n$ for all $n$, and **decreasing** if $a_{n+1} \leq a_n$ for all $n$. It is **monotone** if it is increasing or decreasing.
- The **Monotone Convergence Theorem**: every bounded monotone sequence is convergent.
- $K = \sup A$ if and only if $K$ is an upper bound on $A$ and there is a sequence in $A$ converging to $K$.
- $K = \inf A$ if and only if $K$ is a lower bound on $A$ and there is a sequence in $A$ converging to $K$.

## 3.5 Tutorial problems

1. Use the limit theorems in this chapter to prove that the following sequences converge.

$$\text{(a) } a_n = \frac{7n^2 + 2}{2n^2 + 11n}; \quad \text{(b) } b_n = \frac{(-1)^n}{n^2}; \quad \text{(c) } c_n = \frac{n}{2^n}.$$

2. (a) Let $a_n$ be bounded and $b_n \to 0$. Prove that $a_n b_n \to 0$.
   (b) Let $b_n \to 0$ and $a_n b_n \to 0$. Does it follow that $a_n$ is bounded?
   (c) Let $a_n$ be bounded but not convergent and $a_n b_n \to 0$. Does it follow that $b_n \to 0$?

## 3.6 Homework problems

1. Use the limit theorems in this chapter to prove that the following sequences converge.

$$a_n = \frac{15n^2 - 6}{5n^2 - 1}; \qquad b_n = \frac{\sin(2n + 1)}{n};$$

$$c_n = n - \left(\frac{n+1}{n+2}\right)n; \quad d_n = \frac{9^n}{(n+8)!}.$$

2. Let $a_n$ be the sequence defined inductively by $a_1 = 2$ and

$$a_{n+1} = \frac{1}{2}\left(a_n + \frac{2}{a_n}\right).$$

   (a) Prove by induction that $a_n \in [1, 2]$ for all $n \in \mathbb{Z}^+$.
   (b) Prove that $a_n^2 \geq 2$ for all $n \in \mathbb{Z}^+$.
   (c) Hence prove that the sequence is decreasing.
   (d) We already know that $(a_n)$ is bounded below (by 1) so it follows, by the Monotone Convergence Theorem, that $(a_n)$ converges to some limit $L$. Show that $L > 0$ and $L^2 = 2$.

   *Remark: you have now finally proved that there is a positive real number whose square is 2, that is, you've proved that $\sqrt{2}$ exists! Since you already know that no rational number has square 2, it follows that you now know your first "explicit" irrational number.*

3. (a) Let $(a_n)$ be a sequence such that $a_n \geq K$ for all $n \in \mathbb{Z}^+$ and $a_n \to L$. Prove that $L \geq K$.

(b) Let $(a_n)$ be a sequence such that $a_n \leq M$ for all $n \in \mathbb{Z}^+$ and $a_n \to L$. Prove that $L \leq M$. (Hint: you could define $b_n = -a_n$ and use Theorem 3.5, for example.)

*Remark: taken together, your arguments provide a proof of Proposition 3.4.*

# Chapter 4

# Subsequences

## 4.1 Definition and convergence properties

Informally, a subsequence of a sequence $(a_n)$ is the sequence remaining after we have deleted some (possibly infinitely many) terms from $(a_n)$. To be more precise:

**Definition 4.1.** A sequence $(b_k)$ is a **subsequence** of a sequence $(a_n)$ if there exists a strictly increasing sequence of positive integers $(n_k)$ such that $b_k = a_{n_k}$ for all $k \in \mathbb{Z}^+$.

So all the terms of $(b_k)$ must occur in $(a_n)$ and they must occur in the same order, but we are allowed to skip some terms in $(a_n)$.

**Example 4.2.**

(i) The constant sequence $(b_k) = (1, 1, 1, 1, \ldots)$ is a subsequence of $a_n = (-1)^n$, with $n_k = 2k$. It's the subsequence consisting of only the even-numbered terms of $(a_n)$. From this example we see that a divergent sequence can have a convergent subsequence. Note that we can equally well think of $b_k$ as $a_{2^k}$: the indexing sequence $(n_k)$ is not necessarily unique.

(ii) $b_k = \dfrac{1}{2^k}$ is a subsequence of $a_n = 1/n$, with $n_k = 2^k$. In this case, both $(a_n)$ and $(b_k)$ converge, and their limits are equal.

(iii) $b_k = \dfrac{1}{|3k - 11|}$ is *not* a subsequence of $a_n = 1/n$ even though every term of $(b_k)$ is the reciprocal of a positive integer, so appears somewhere in the sequence $(a_n)$. The problem is that the terms don't occur in the right order:

$$(b_k) = \left(\frac{1}{8}, \frac{1}{5}, \frac{1}{2}, 1, \frac{1}{4}, \frac{1}{7}, \ldots\right)$$

$$= (a_8, a_5, a_2, a_1, a_4, a_7, \ldots)$$

(iv) Every sequence $(a_n)$ is a subsequence of itself! We take $n_k$ to be the increasing sequence $n_k = k$.

(v) Given any sequence $(a_n)$, and any positive integer $p$, $b_k = a_{k+p}$ is a subsequence of $(a_n)$, with $n_k = k + p$. This is just the subsequence obtained by omitting the first $p$ terms of $(a_n)$. We call such a subsequence a **tail** of $(a_n)$.  □

As we have seen, it is possible for a divergent sequence to have a convergent subsequence. On the other hand, if a sequence is convergent, with limit $L$, one expects that all its subsequences should also converge to $L$. This turns out to be true.

**Theorem 4.3.** *If $a_n \to L$ and $(b_k)$ is a subsequence of $(a_n)$, then $b_k \to L$.*

**Proof.** Let $\varepsilon > 0$ be given. Since $a_n \to L$, there exists $N \in \mathbb{Z}^+$ such that $|a_n - L| < \varepsilon$ for all $n \geq N$. Since $(b_k)$ is a subsequence of $(a_n)$, $b_k = a_{n_k}$ where $n_k$ is a strictly increasing sequence of positive integers. Now $n_1 \geq 1$, and if $n_k \geq k$ then $n_{k+1} \geq n_k + 1 \geq k + 1$. Hence, by induction, $n_k \geq k$ for all $k$. Hence, for all $k \geq N$, $n_k \geq n_N \geq N$, so $|b_k - L| = |a_{n_k} - L| < \varepsilon$. Hence $(b_k) \to L$.  □

We can use this to give a sneaky proof that $a_n = (-1)^n$ doesn't converge, without having to resort to an $\varepsilon$–$N$ argument:

**Example 2.8 (revisited).** Claim: $a_n = (-1)^n$ doesn't converge to any limit.

**Proof.** Assume, to the contrary, that $a_n \to L$. Then, by Theorem 4.3, the subsequences $a_{2k}$ and $a_{2k+1}$ also converge to $L$. But $a_{2k} = 1 \to 1$ and $a_{2k+1} = -1 \to -1$, so $1 = L = -1$, a contradiction.  □

Note that we've actually used Theorem 4.3 a couple of times already. In Examples 3.15 and 2.3 (revisited) we asserted that if $a_n \to L$ then $b_n = a_{n+1} \to L$ also (see page 39). This follows from Theorem 4.3 since $(b_n)$

is a subsequence of $(a_n)$. It's important to realize that Theorem 4.3 does *not* say that if a subsequence of $(a_n)$ converges to $L$ then $(a_n)$ converges to $L$ (we've already seen a counterexample to this: $a_n = (-1)^n$). Just showing that a subsequence of $(a_n)$ converges is not enough (usually) to show that $(a_n)$ converges. There is an important case, however, where convergence of the subsequence does imply convergence of the whole sequence, namely, any tail subsequence, that is, a subsequence obtained by chopping off the first $m$ terms of the sequence.

**Lemma 4.4 (Tail Lemma).** *Let $(a_n)$ be a real sequence, $m$ be a positive integer and $b_n = a_{n+m}$. If $b_n \to L$ then $a_n \to L$.*

**Proof.** Assume $b_n \to L$. Let $\varepsilon > 0$ be given. Then there exists $N_1 \in \mathbb{Z}^+$ such that, for all $n \geq N_1$, $|b_n - L| < \varepsilon$. Let $N = N_1 + m$. Then, for all $n \geq N$, $n - m \geq N_1$, so

$$|a_n - L| = |b_{n-m} - L| < \varepsilon.$$

Hence $a_n \to L$. $\qquad\qquad\qquad\qquad\qquad\qquad\qquad\qquad\qquad\qquad\square$

This lemma is very useful in practice because it allows us to ignore any number of terms at the start of a sequence when attempting to prove convergence.

**Example 4.5.** Claim $a_n = \dfrac{n^3}{2^n}$ converges.
This isn't immediately obvious, because $a_n$ is a quotient of functions of $n$ both of which grow unbounded as $n$ gets large, so it's not obvious whether $(a_n)$ is even bounded, let alone convergent. To prove convergence, we'd like to use the Monotone Convergence Theorem, but there's a problem: $(a_n)$ is neither increasing nor decreasing! In fact

$$a_1 < a_2 < a_3 < a_4,$$

but $a_n \geq a_{n+1}$ from there on (i.e. for all $n \geq 4$). So although $(a_n)$ isn't decreasing, it is "eventually" decreasing, and by Lemma 4.4, that's good enough.

**Proof.** First note that $a_n$ is bounded below (by 0, for example). Now

$$\frac{a_{n+1}}{a_n} = \frac{(n+1)^3}{2n^3} = \frac{n^3 + 3n^2 + 3n + 1}{2n^3}$$

$$\leq \frac{n^3 + 3n^2 + 3n^2 + n^2}{2n^3} \qquad \text{(since } n \geq 1\text{)}$$

$$= \frac{1}{2} + \frac{7}{2n} < 1 \qquad \text{provided } n > 7.$$

Hence, for all $n \geq 8$, $a_{n+1} < a_n$. Hence $b_n = a_{n+7}$ is a decreasing sequence, bounded below. By the Monotone Convergence Theorem, $(b_n)$ converges so, by Lemma 4.4, $(a_n)$ converges also. $\qquad \square$

The above argument does *not* tell us what the limit of $(a_n)$ is, only that the limit exists. In fact, $a_n \to 0$, but to show this we must work harder. We'll return to this later, after we've studied convergence of series.

We next identify a class of *pairs* of subsequences whose convergence to a common limit implies convergence of their parent sequence.

**Definition 4.6.** A pair of subsequences $(a_{n_k})$, $(a_{m_k})$ is a **covering pair** for $(a_n)$ if, for each $n \in \mathbb{Z}^+$, there exists $k \in \mathbb{Z}^+$ such that $n = n_k$ or $n = m_k$. Hence every term in the sequence $(a_n)$ occurs in (at least) one of the subsequences $(a_{n_k})$, $(a_{m_k})$.

**Example 4.7.** For any sequence $(a_n)$, the odd and even numbered subsequences $(a_{2k-1})$, $(a_{2k})$ are a covering pair.

**Lemma 4.8.** Let $(a_{n_k})$, $(a_{m_k})$ be a covering pair for $(a_n)$. If $a_{n_k} \to L$ and $a_{m_k} \to L$ then $a_n \to L$.

**Proof.** Let $\varepsilon > 0$ be given. Since $a_{n_k} \to L$, there exists $K_1 \in \mathbb{Z}^+$ such that, for all $k \geq K_1$, $|a_{n_k} - L| < \varepsilon$. Similarly, since $a_{m_k} \to L$, there exists $K_2 \in \mathbb{Z}^+$ such that, for all $k \geq K_2$, $|a_{m_k} - L| < \varepsilon$. Let $N = \max\{n_{K_1}, m_{K_2}\}$, and consider any $n \geq N$. By the definition of a covering pair, either $n = n_k$ or $n = m_k$, for some $k \in \mathbb{Z}^+$. If $n = n_k$, then $k \geq K_1$, since $n \geq n_{K_1}$ and $(n_k)$ is strictly increasing. Hence,

$$|a_n - L| = |a_{n_k} - L| < \varepsilon.$$

Similarly, if $n = m_k$, then $k \geq K_2$, since $n \geq m_{K_2}$ and $(m_k)$ is strictly increasing. Hence,

$$|a_n - L| = |a_{m_k} - L| < \varepsilon.$$

We conclude that $|a_n - L| < \varepsilon$ for all $n \geq N$, and so $a_n \to L$. $\qquad \square$

**Example 4.9.** Claim: $a_n = (1 + (-1)^n)/n \to 0$.

**Proof.** The odd and even subsequences of $(a_n)$ are

$$a_{2k-1} = 0,$$
$$a_{2k} = \frac{1}{k}.$$

These form a covering pair, and we have already shown that they both converge to 0. Hence, by Lemma 4.8, $a_n \to 0$. $\qquad\square$

## 4.2 The Bolzano–Weierstrass Theorem

We have seen that all subsequences of a convergent sequence are convergent and that non-convergent sequences can sometimes have convergent subsequences. The goal of this section is to prove a famous and very powerful result called the Bolzano–Weierstrass[1] Theorem, which says that every *bounded* sequence has at least one convergent subsequence. It won't be immediately clear why this fact is such a big deal. Later, we will use it to prove the Extreme Value Theorem, which is a crucial result when one comes to study calculus. (For a reminder of what the Extreme Value Theorem says, refer to page xii.)

The strategy of proof is quite straightforward: we first prove that *every* sequence has a monotone subsequence. Having done this, the Bolzano-Weierstrass Theorem follows immediately from the Monotone Convergence Theorem.

**Lemma 4.10.** *Every sequence has a monotone subsequence.*

**Proof.** Let the sequence in question be $(a_n)$. We make the following definition: a term $a_m$ is **dominant** if $a_n \le a_m$ for all $n > m$ (that is, $a_m$ is at least as big as all subsequent terms). Consider the set of all dominant terms of the sequence $(a_n)$. There are exactly two possibilities: either (i) the set of dominant terms is infinite, or (ii) the set of dominant terms is finite (which includes the case where there are no dominant terms).

Case (i): if the set of dominant terms is infinite, then the subsequence consisting of just these terms is, by definition, decreasing. Hence $(a_n)$ has a decreasing subsequence.

Case (ii): if the set of dominant terms is finite (or empty), there is a term in the sequence, $a_m$ say, beyond which there are no dominant terms.

---

[1] Weierstraß for German purists.

Let $n_1 = m + 1$. Since $a_{n_1}$ is *not* dominant, there exists $n > n_1$ such that $a_n > a_{n_1}$. Choose such an $n$ and call it $n_2$. Then $a_{n_2}$ is also not dominant (since $n_2 > n_1 > m$), so there exists a positive integer greater than $n_2$, let's call it $n_3$, such that $a_{n_3} > a_{n_2}$. In this way we construct a strictly increasing sequence of positive integers $n_1, n_2, n_3, \ldots$ such that $a_{n_{k+1}} > a_{n_k}$ for all $k$. But then $(a_{n_k})$ is an increasing subsequence of $(a_n)$. Hence $(a_n)$ has an increasing subsequence.

In either case, we see that $(a_n)$ has a monotone subsequence, as was to be proved. □

**Theorem 4.11 (The Bolzano–Weierstrass Theorem).** *Let $(a_n)$ be a bounded real sequence. Then $(a_n)$ has a convergent subsequence.*

**Proof.** By Lemma 4.10, $(a_n)$ has a monotone subsequence $(a_{n_k})$. But the range of $(a_{n_k})$ is a subset of the range of $(a_n)$, so $(a_{n_k})$ is also bounded (by the same upper and lower bounds as $(a_n)$), and hence converges by the Monotone Convergence Theorem. □

Of course, the converse of this Theorem is false: just because a sequence has a convergent subsequence, it doesn't follow that it's bounded. For example

$$a_n = \begin{cases} n & n \text{ odd} \\ 0 & n \text{ even} \end{cases}$$

has a convergent subseqence, $a_{2k} = 0 \to 0$, but is clearly unbounded above.

**Example 2.2 (revisited).** Recall we defined the sequence $a_n = \sin n$ and noted that its terms "bounce around" in the interval $(-1, 1)$ seemingly at random. By the Bolzano–Weierstrass Theorem we know that $a_n$ has at least one convergent subsequence, although we have absolutely no idea how to write it down, or what its limit might be. (In fact, given any $L \in [-1, 1]$ it turns out that this sequence has a subsequence converging to $L$, but proving this goes somewhat beyond the scope of this book.) □

## 4.3 Summary

- A **subsequence** of a sequence $(a_n)$ is any sequence of the form $(a_{n_k})$ where $n_k$ is a strictly increasing sequence of positive integers.
- Non-convergent sequences can have convergent subsequences.
- If $(a_n)$ converges to $L$, every subsequence of $(a_n)$ also converges to $L$.
- A **tail** of a sequence $(a_n)$ is a subsequence of the form $(a_{m+n})$ for some constant $m \in \mathbb{Z}^+$.
- If a sequence has a convergent tail then the sequence itself converges (to the same limit).
- A **covering pair** of subsequences of $(a_n)$ is any pair $(a_{n_k})$, $(a_{m_k})$ such that $\{n_k : k \in \mathbb{Z}^+\} \cup \{m_k : k \in \mathbb{Z}^+\} = \mathbb{Z}^+$.
- If both subsequences in a covering pair converge to the same limit, then the sequence itself converges (to this common limit).
- Every sequence has a monotone subsequence. Together with the Monotone Convergence Theorem, this implies:
- The **Bolzano–Weierstrass Theorem**: every *bounded* sequence has a convergent subsequence.

## 4.4   Tutorial problems

1. For each of the following sequences,

   (i) Write down the dominant terms of the sequence.
   (ii) Construct a monotone subsequence of the sequence.
   (iii) Determine whether the sequence has a convergent subsequence.

$$\text{(a) } a_n = \frac{1-(-1)^n}{2n-9}, \qquad \text{(b) } b_n = ((-1)^n - 2)n.$$

2. Let $(a_n)$ be an unbounded sequence. Show that it has a subsequence $(a_{n_k})$ such that $1/a_{n_k} \to 0$.

3. Let $x_1 = 1$, and for all $n \geq 1$ let

$$x_{n+1} = 1 + \frac{1}{x_n}.$$

   (i) Show, by induction, that for all $n$ odd,
   $$1 + x_n - x_n^2 > 0,$$
   and for all $n$ even,
   $$1 + x_n - x_n^2 < 0.$$

   (ii) Deduce that the subsequence $(x_{2k-1})$ is increasing and the subsequence $(x_{2k})$ is decreasing.
   (iii) Deduce that both $(x_{2k-1})$ and $(x_{2k})$ converge.
   (iv) Show that $(x_n)$ converges.

## 4.5   Homework problems

1. (a) Let $a_n \to 2$. Prove from first principles (i.e. give a direct $\varepsilon$–$N$ proof) that $a_n^2 \to 4$.
   (b) Let $a_n^2 \to 4$. Prove that $(a_n)$ has a subsequence which converges either to 2 or $-2$.
   (c) Let $a_n^2 \to 4$. Prove from first principles (i.e. give a direct $\varepsilon$–$N$ proof) that $|a_n| \to 2$.

2. For each of the following sequences,

   (i) Write down the dominant terms of the sequence.
   (ii) Construct a monotone subsequence of the sequence.
   (iii) Determine whether the sequence has a convergent subsequence.

$$a_n = \begin{cases} -n & \text{if } n \text{ divisible by 3} \\ (-1)^n & \text{otherwise} \end{cases}, \qquad b_n = \frac{1}{n^2 - 13}, \qquad c_n = (-1)^n n.$$

3. Let $x_1$ be some real number other than 0 or 1 and, for all $n \geq 1$, let

$$x_{n+1} = 1 - \frac{1}{x_n}.$$

(a) Verify that the sequence $(x_n)$ is well-defined.

(b) Show that $(x_n)$ diverges.

(c) Show that the subsequence $(x_{3k})$ converges.

# Chapter 5

# Series

## 5.1 Definition and convergence

Informally, a series is an infinite sum

$$\sum_{n=1}^{\infty} a_n = a_1 + a_2 + a_3 + \cdots,$$

for example

$$\sum_{n=1}^{\infty} \frac{1}{n} = 1 + \frac{1}{2} + \frac{1}{3} + \frac{1}{4} + \cdots.$$

Before proceeding any further, we should ask ourselves what, exactly, the expression on the right-hand side of these equations really means. In fact, it turns out that a series is really nothing more than a sequence – but *not* the sequence $(a_n)$, rather the sequence of *partial sums*.

**Definition 5.1.** Given a **series** $\sum_{n=1}^{\infty} a_n$ we define $a_n$ to be the $n^{\text{th}}$ **term** of the series and

$$s_k = \sum_{n=1}^{k} a_n$$

to be its $k^{\text{th}}$ **partial sum**. The series **converges** if the sequence $(s_k)$ converges in the usual sense (Definition 2.5). In this case we (extremely confusingly) use the same symbol, $\sum_{n=1}^{\infty} a_n$, to denote its limit.[1] If $(s_k)$ does not converge we say the series **diverges**.

**Remark.** Do not confuse the $k^{\text{th}}$ *term* of a series, $a_k$, with its $k^{\text{th}}$ *partial sum*, $s_k$. In particular, convergence of the series means convergence of $(s_k)$, *not* convergence of $(a_k)$.

---

[1]Don't blame me: this convention is too widespread to challenge.

**Example 5.2 (Harmonic series).** Claim: the series $\displaystyle\sum_{n=1}^{\infty}\frac{1}{n}$ diverges.

**Proof.** We will show that the sequence of partial sums $(s_k)$ has an unbounded, and hence divergent, subsequence. It then follows from Theorem 4.3 that $(s_k)$ itself diverges. Consider $s_{2^p}$:

$$s_{2^p} = 1 + \frac{1}{2} + \left(\frac{1}{3}+\frac{1}{4}\right) + \left(\frac{1}{5}+\frac{1}{6}+\frac{1}{7}+\frac{1}{8}\right) + \cdots + \left(\frac{1}{2^{p-1}+1}+\cdots+\frac{1}{2^p}\right)$$

$$> 1 + \frac{1}{2} + \left(\frac{1}{4}+\frac{1}{4}\right) + \left(\frac{1}{8}+\frac{1}{8}+\frac{1}{8}+\frac{1}{8}\right) + \cdots + \left(\frac{1}{2^p}+\cdots+\frac{1}{2^p}\right)$$

$$= 1 + \frac{1}{2} + \frac{1}{2} + \frac{1}{2} + \cdots + \frac{1}{2}$$

$$= 1 + \frac{p}{2}.$$

This is unbounded, so $(s_k)$ is divergent. $\qquad\square$

**Remark.** This is the most famous of all divergent series, called the *harmonic series*. Note that the *terms* of the series $a_n = 1/n$ converge to 0. Nonetheless, the series itself does not converge.

**Example 5.3.** Claim: the series $\displaystyle\sum_{n=1}^{\infty}\frac{1}{n(n+1)}$ converges to 1.

**Proof.** Note that

$$a_n = \frac{1}{n(n+1)} = \frac{1}{n} - \frac{1}{n+1}$$

so

$$s_k = \sum_{n=1}^{k}\left(\frac{1}{n}-\frac{1}{n+1}\right) = \sum_{n=1}^{k}\frac{1}{n} - \sum_{n=1}^{k}\frac{1}{n+1}$$

$$= \sum_{n=1}^{k}\frac{1}{n} - \sum_{m=2}^{k+1}\frac{1}{m} \qquad (\text{where } m = n+1)$$

$$= 1 - \frac{1}{k+1}.$$

Hence $s_k \to 1$. $\qquad\square$

    This particular example was cooked up so that an explicit expression for the partial sum $s_k$ could be found. The key property of $s_k$ allowing us to do this is that it is a **telescoping sum**: it takes the form

$$s_k = \sum_{n=1}^{k} (b_{n+1} - b_n).$$

(In this case $b_n = -1/n$.) For any such sum

$$s_k = \sum_{n=1}^{k} b_{n+1} - \sum_{n=1}^{k} b_n = \sum_{m=2}^{k+1} b_m - \sum_{n=1}^{k} b_n = b_{k+1} - b_1$$

since all but the last term in the first sum and the first term in the second sum cancel, a bit like an old-fashioned telescope collapsing for storage.

Another important example where we can compute the limit exactly is the *geometric series*:

**Example 5.4 (Geometric series).** Let $\alpha \in (-1, 1)$ and consider the series

$$\sum_{n=0}^{\infty} \alpha^n = 1 + \alpha + \alpha^2 + \cdots.$$

This is the **geometric series with common ratio** $\alpha$. We claim that it converges to $\dfrac{1}{1-\alpha}$.

**Proof.**

$$s_k = 1 + \alpha + \alpha^2 + \cdots + \alpha^k$$
$$\Rightarrow \quad \alpha s_k = \alpha + \alpha^2 + \alpha^3 + \cdots + \alpha^{k+1}$$
$$\Rightarrow \quad (1 - \alpha)s_k = 1 - \alpha^{k+1}$$
$$\Rightarrow \quad s_k = \frac{1 - \alpha^{k+1}}{1 - \alpha}.$$

Now $\alpha^{k+1} \to 0$ (see Example 3.15), so $s_k \to 1/(1-\alpha)$. $\square$

Usually it's not possible to find a simple formula for $s_k$, so we need to develop ways of determining whether a series converges by dealing directly with its *terms*, $a_n$, rather than its *partial sums*, $s_k$. In the next section we will construct several convergence tests for this purpose. It's important to realize that these tests will only tell us *whether* a series converges, not *what limit* it converges to (if it does converge). For example, we will be able to prove that

$$\sum_{n=1}^{\infty} \frac{1}{n^2}$$

converges, even though we have absolutely no idea what it converges to.[2]

---

[2] Actually, it converges to $\pi^2/6$, but to prove this requires some very fancy stuff.

## 5.2   Convergence tests for series

Recall that a series $\sum_{n=1}^{\infty} a_n$ converges if and only if its sequence of partial sums, $s_k = \sum_{n=1}^{k} a_n$ converges. Our first convergence test gives a very simple *necessary* but *not sufficient* condition for convergence.

**Theorem 5.5 (The Divergence Test).** *If* $\sum_{n=1}^{\infty} a_n$ *converges, then* $(a_n)$ *converges to* 0.

**Proof.** By assumption, $s_k = \sum_{n=1}^{k} a_n$ converges to some limit $L$. Hence, $s_{k+1} \to L$ also by Theorem 4.3 (it's a subsequence of $(s_k)$). Hence $a_{k+1} = s_{k+1} - s_k \to L - L = 0$ by the Algebra of Limits, so the sequence $a_n \to 0$ by the Tail Lemma (Lemma 4.4).                                              $\square$

**Warning!** This theorem says that *if* $\sum a_n$ converges, then $a_n \to 0$. It does *not* say that if $a_n \to 0$ then $\sum a_n$ converges. We have already seen a counterexample, the harmonic series: $a_n = 1/n \to 0$ but $\sum_{n=1}^{\infty} \frac{1}{n}$ does *not* converge!

**Example 5.6.** $\sum_{n=1}^{\infty} \frac{n^2-1}{n^2+1}$ does not converge, by the Divergence Test (since its terms $a_n$ converge to 1, not 0).

If the terms $a_n$ of a series are positive, then the sequence of partial sums is increasing, since $s_{k+1} = s_k + a_{k+1} > s_k$. Hence, such a series is convergent if (and only if) its sequence of partial sums is *bounded*, by the Monotone Convergence Theorem. This observation allows us to formulate several useful tests which apply in the special case $a_n > 0$ for all $n$.

**Theorem 5.7 (The Comparison Test).** *Let* $a_n > 0$ *and* $b_n > 0$ *for all* $n$.

(i) *If the sequence* $\dfrac{a_n}{b_n}$ *is bounded above and* $\sum_{n=1}^{\infty} b_n$ *converges, then* $\sum_{n=1}^{\infty} a_n$ *converges.*

(ii) *If the sequence* $\dfrac{b_n}{a_n}$ *is bounded above and* $\sum_{n=1}^{\infty} b_n$ *diverges, then* $\sum_{n=1}^{\infty} a_n$ *diverges.*

**Proof.** Let $s_k = \sum_{n=1}^{k} a_n$ and $t_k = \sum_{n=1}^{k} b_n$, and note that both these sequences are increasing.

(i) Since $a_n/b_n$ is bounded above, there exists $K \in \mathbb{R}$ such that $(a_n/b_n) \le K$ for all $n$, and hence, $a_n \le K b_n$ for all $n$ . Hence $s_k \le K t_k$.

But the sequence $(t_k)$, by assumption, converges, so is itself bounded above, by $L$ say. Hence $s_k \leq KL$. Since $(s_k)$ is increasing and bounded above, it converges by the Monotone Convergence Theorem.

(ii) Since $b_n/a_n$ is bounded above, there exists $K \in \mathbb{R}$ such that $(b_n/a_n) \leq K$ for all $n$, and hence $b_n \leq K a_n$ for all $n$. Clearly $K > 0$. Hence $s_k \geq t_k/K$. But the sequence $(t_k)$ is increasing and not convergent, so must be unbounded above (or else we'd have a counterexample to the Monotone Convergence Theorem), and hence $(s_k)$ is also unbounded above. Hence $(s_k)$ diverges. $\qquad\square$

We can use this theorem to show that a series of interest converges/diverges by comparing it to a series which we already know converges/diverges.

**Example 5.8.** Claim: $\displaystyle\sum_{n=1}^{\infty} \frac{1}{n^2}$ converges.

**Proof.** Let $a_n = \dfrac{1}{n^2}$ and $b_n = \dfrac{1}{n(n+1)}$. Then, as shown in Example 5.3, $\sum_{n=1}^{\infty} b_n$ converges. Now

$$\frac{a_n}{b_n} = \frac{1}{n^2} \times n(n+1) = 1 + \frac{1}{n} \to 1.$$

Since $a_n/b_n$ converges, it is bounded (Proposition 3.3), so by part (i) of the Comparison Test, $\sum_{n=1}^{\infty} a_n$ also converges. $\qquad\square$

**Remark.** Note that we didn't show that $a_n/b_n$ was bounded directly. Instead, we used the Algebra of Limits to show that it converges. Every convergent sequence is bounded, so it follows that $a_n/b_n$ is bounded, as the Comparison Test requires.

**Example 5.9.** Let $p \in \mathbb{Z}^+$ and $p > 2$. Then $\displaystyle\sum_{n=1}^{\infty} \frac{1}{n^p}$ converges.

**Proof.** Let $a_n = 1/n^p$ and $b_n = 1/n^2$. Then $\sum_{n=1}^{\infty} b_n$ converges, by Example 5.8. Now

$$\frac{a_n}{b_n} = \frac{1}{n^p} \times n^2 = \frac{1}{n^{p-2}} \to 0$$

since $p > 2$. Since $a_n/b_n$ converges, it is bounded, so by part (i) of the Comparison Test, $\sum_{n=1}^{\infty} a_n$ also converges. $\qquad\square$

**Example 5.10.** $\displaystyle\sum_{n=1}^{\infty} \frac{n}{2n^2 + \sin n}$ diverges.

**Proof.** Let $a_n = n/(2n^2 + \sin n)$ and $b_n = 1/n$. Then $\sum_{n=1}^{\infty} b_n$ diverges, by Example 5.2. Now

$$\frac{b_n}{a_n} = \frac{2n^2 + \sin n}{n^2} = 2 + \frac{\sin n}{n^2}.$$

But $\quad -\dfrac{1}{n^2} \leq \dfrac{\sin n}{n^2} \leq \dfrac{1}{n^2} \quad \Rightarrow \quad \dfrac{\sin n}{n^2} \to 0 \quad$ (by the Squeeze Rule)

so $b_n/a_n \to 2$. Since $b_n/a_n$ converges, it is bounded, so by part (ii) of the Comparison Test, $\sum_{n=1}^{\infty} a_n$ also diverges. $\qquad\square$

To use the Comparison Test, you have to have some intuition about whether the series converges. If you suspect it does, compare it with a series you know converges. If you suspect it doesn't, compare it with a series you know doesn't. In Example 5.10 I recognized that, for large $n$, the $\sin n$ part is essentially irrelevant, so the terms look roughly like $n/(2n^2) = 1/(2n)$. So I compared the series to the divergent series $\sum(1/n)$.

Our next convergence test (everyone's favourite) involves only the terms of the series under consideration and is very easy to apply. Its only disadvantage is that it very often gives an inconclusive answer.

**Theorem 5.11 (The Ratio Test).** *Let $a_n > 0$ for all $n$ and assume that the sequence of ratios of consecutive terms converges, that is, $\dfrac{a_{n+1}}{a_n} \to L$ for some $L \in \mathbb{R}$.*

*(i) If $L < 1$ then $\sum_{n=1}^{\infty} a_n$ converges.*
*(ii) If $L > 1$ then $\sum_{n=1}^{\infty} a_n$ diverges.*

**Proof.** As usual, let $s_k = \sum_{n=1}^{k} a_n$, and note that, since $a_n > 0$, the sequence $(s_k)$ is increasing and so converges if and only if it is bounded above.

(i) Let $r_n = a_{n+1}/a_n$. Choose any $\gamma \in (L, 1)$ and let $\varepsilon = \gamma - L > 0$. Then since $r_n \to L$, there exists $N \in \mathbb{Z}^+$ such that, for all $n \geq N$, $|r_n - L| < \varepsilon$, so $r_n < L + \varepsilon = \gamma$. Hence, for all $n \geq N$, $a_{n+1} < \gamma a_n$. Let

$$b_n = \gamma^{n-N} a_N.$$

Then $b_n \geq a_n$ for all $n \geq N$. (Check this by induction: $b_N = a_N$, and if $b_n \geq a_n$ then $b_{n+1} = \gamma b_n \geq \gamma a_n > a_{n+1}$.) Hence, for all $k \geq N$,

$$s_k = \sum_{n=1}^{k} a_n = s_{N-1} + \sum_{n=N}^{k} a_n$$

$$\leq s_{N-1} + \sum_{n=N}^{k} b_n$$

$$< s_{N-1} + \sum_{n=1}^{k} b_n$$

$$= s_{N-1} + \frac{a_N}{\gamma^N} \sum_{n=1}^{k} \gamma^n =: t_k.$$

Now $(t_k)$ converges by Example 5.4 (up to constants, it's the geometric series of common ratio $\gamma \in (0,1)$), so $(t_k)$ is bounded above, and hence $(s_k)$ is bounded above. Hence $(s_k)$ converges.

(ii) Let $r_n = a_{n+1}/a_n$. Choose any $\gamma \in (1, L)$ and let $\varepsilon = L - \gamma > 0$. Then, since $r_n \to L$, there exists $N \in \mathbb{Z}^+$ such that, for all $n \geq N$, $|r_n - L| < \varepsilon$, so $r_n > L - \varepsilon = \gamma$. Hence, for all $n \geq N$, $a_{n+1} > \gamma a_n > a_n$. Hence the subsequence $(a_N, a_{N+1}, a_{N+2}, \ldots)$ is bounded below by $a_N > 0$. Hence this subsequence does not converge to 0 (Proposition 3.4) so the sequence $(a_n)$ does not converge to 0 (Theorem 4.3), so the series $\sum_{n=1}^{\infty} a_n$ diverges, by the Divergence Test (Theorem 5.5). $\qquad \square$

The ratio test is very handy but it gives no useful information in the case $L = 1$. For example, if you try to use it for $\sum_{n=1}^{\infty} n^{-p}$ it will be inconclusive.

**Example 5.12.** Claim: the series $\sum_{n=0}^{\infty} \frac{1}{n!}$ converges.

**Proof.** Let $a_n = 1/n!$. Then $a_{n+1}/a_n = n!/(n+1)! = 1/(n+1) \to 0 < 1$, so the series converges by the Ratio Test. $\qquad \square$

The limit of this series is a very famous number, called **Euler's number** $e$. Note that the series starts with $n = 0$, and recall that $0! = 1$. A famous fact about $e$ is that it is *irrational*.

**Proof.** Assume, to the contrary, that $e = p/q$ where $p, q \in \mathbb{Z}^+$, and denote, as usual, the $k^{\text{th}}$ partial sum by $s_k$. Then $q!e$ is a positive integer, as is

$$q!s_q = q! \sum_{n=0}^{q} \frac{1}{n!} = q! + q! + [q(q-1)\cdots 3] + [q(q-1)\cdots 4] + \cdots + [1].$$

Hence, their difference, $q!(e - s_q)$, is also an integer and must be positive, since $e$ is strictly bigger than any of its partial sums. By definition, $q!(e - s_q)$ is the limit of the convergent sequence

$$t_k = q! \sum_{n=q+1}^{q+k} \frac{1}{n!}$$

$$= \frac{1}{q+1} + \frac{1}{(q+1)(q+2)} + \cdots + \frac{1}{(q+1)(q+2)\cdots(q+k)}.$$

Compare this with the partial sum

$$u_k = \frac{1}{q+1} + \frac{1}{(q+1)^2} + \cdots + \frac{1}{(q+1)^k}$$

$$= \frac{1}{q+1} \sum_{n=0}^{k-1} \frac{1}{(q+1)^n}.$$

As we have seen (Example 5.4),

$$u_k \to \frac{1}{q+1} \left( \frac{1}{1 - 1/(q+1)} \right) = \frac{1}{q}.$$

But, for all $k \geq 2$,

$$u_k - t_k \geq \frac{1}{(q+1)^2} - \frac{1}{(q+1)(q+2)} = \frac{1}{(q+1)^2(q+2)}.$$

Hence, by Proposition 3.4,

$$\lim t_k \leq \lim u_k - \frac{1}{(q+1)^2(q+2)} < \frac{1}{q} \leq 1.$$

That is, the positive integer $q!(e - s_q)$ is strictly less than 1, a contradiction.

$\square$

Another way to define $e$ is as the limit of the *sequence* $(1 + 1/n)^n$. However, proving that this sequence converges is much trickier than proving that the series $\sum(1/n!)$ converges, and it's also not so obvious from the sequence definition that $e$ is irrational. We will return to this later (Proposition 12.6 on page 198).

$\square$

In the next example we give a very sneaky use of the Ratio Test: we use it to prove that a certain *sequence* converges to 0 by showing that the series whose *terms* are the sequence converges! Recall that in Example 4.5 we used the Monotone Convergence Theorem to show that

$$a_n = \frac{n^3}{2^n}$$

converged. This is not entirely obvious because both the numerator and the denominator increase without bound as $n$ gets large, so it's not clear whether the polynomial in the numerator wins (so $a_n$ diverges), the exponential factor in the denominator wins (so $a_n \to 0$) or the two reach a draw (so $a_n \to L > 0$). While our previous argument showed the sequence converges, so the polynomial doesn't win, it didn't tell us what the limit is. In fact, the limit is 0. This is an example of a more general fact: exponential growth always beats polynomial growth. To deal with Example 4.5 just choose $p = 3$ and $\alpha = 1/2$ in the following example:

**Example 5.13 (Exponentials beat polynomials).** Let $p \in \mathbb{Z}^+$ and $\alpha \in (0, 1)$ be constant. Then $a_n = n^p \alpha^n \to 0$.

**Proof.** Consider the series $\sum_{n=1}^{\infty} a_n$. This has

$$\frac{a_{n+1}}{a_n} = \frac{(n+1)^p \alpha^{n+1}}{n^p \alpha^n} = \left(1 + \frac{1}{n}\right)^p \alpha \to \alpha < 1.$$

Hence $\sum_{n=1}^{\infty} a_n$ converges, by the Ratio Test, so $a_n \to 0$ by the Divergence Test. $\square$

## 5.3 Alternating series

Apart from the Divergence Test, all the convergence tests we studied in section 5.2 assumed that the terms of the series are strictly positive. In practice, most series of interest have this property. However, there is another special case which arises quite commonly and which can be handled using a convenient convergence test.

**Definition 5.14.** A real series $\sum_{n=1}^{\infty} b_n$ is said to be **alternating** if $b_n \neq 0$ and $b_{n+1}/b_n < 0$ for all $n$.

The point is that consecutive terms of the series alternate in sign.

**Example 5.15.** $\sum_{n=1}^{\infty}(-1)^n$ is an alternating series. Its partial sums are

$$s_k = \begin{cases} -1 & k \text{ odd} \\ 0 & k \text{ even} \end{cases}.$$

Hence, this series diverges. □

Clearly every alternating series can be written in the form

$$\sum_{n=1}^{\infty}(-1)^{n+1}a_n \quad \text{or} \quad -\sum_{n=1}^{\infty}(-1)^{n+1}a_n$$

where $a_n > 0$ for all $n$. We will state and prove our convergence test for series of the left hand type, but it applies equally well to those of the right hand type (by the Algebra of Limits).

**Theorem 5.16 (The Alternating Series Test).** *Let $(a_n)$ be a decreasing sequence of positive numbers which converges to 0. Then*

$$\sum_{n=1}^{\infty}(-1)^{n+1}a_n$$

*converges.*

**Proof.** As usual, let $s_k$ denote the $k^{\text{th}}$ partial sum of the series. Consider the subsequence of even numbered partial sums $(s_{2m})$. Now

$$s_{2m} = (a_1 - a_2) + (a_3 - a_4) + \cdots + (a_{2m-1} - a_{2m})$$
$$\Rightarrow \quad s_{2m+2} - s_{2m} = a_{2m+1} - a_{2m+2} \geq 0 \quad .$$

so the sequence $(s_{2m})$ is increasing. Further,

$$s_{2m} = a_1 - (a_2 - a_3) - \cdots - (a_{2m-2} - a_{2m-1}) - a_{2m} < a_1$$

so $(s_{2m})$ is bounded above, by $a_1$. Hence, by the Monotone Convergence Theorem, $(s_{2m})$ converges to some limit $L$.

Consider now the subsequence of odd numbered partial sums $(s_{2m-1})$. By the alternating property, $s_{2m} = s_{2m-1} - a_{2m}$, so

$$s_{2m-1} = s_{2m} + a_{2m}.$$

Now $a_n \to 0$ by assumption, so the subsequence $a_{2m} \to 0$ also (Theorem 4.3), and we just showed that $s_{2m} \to L$, so $s_{2m-1} \to L$ also by the Algebra of Limits. But $(s_{2m-1})$, $(s_{2m})$ are a covering pair for $(s_n)$ (see Definition 4.6), so $s_n \to L$ by Lemma 4.8. Hence $(s_n)$ converges. □

**Example 5.17 (The alternating harmonic series).**

$$\sum_{n=1}^{\infty} \frac{(-1)^{n+1}}{n}$$

converges, by the Alternating Series Test, since $a_n = 1/n$ is positive, decreasing, and converges to 0. □

The condition that $(a_n)$ should be *decreasing* is a crucial part of the Alternating Series Test. Without it, the conclusion can be false.

**Example 5.18.** Consider the sequence

$$a_n = \begin{cases} 1/n^2 & n \text{ odd} \\ 1/n & n \text{ even} \end{cases}$$

This is positive and converges to 0 but is *not* decreasing. I claim that the alternating series $\sum_{n=1}^{\infty} (-1)^{n+1} a_n$ diverges.

**Proof.** Consider the even subsequence of partial sums

$$s_{2m} = 1 - \frac{1}{2} + \frac{1}{3^2} - \frac{1}{4} + \cdots + \frac{1}{(2m-1)^2} - \frac{1}{2m}$$

$$= \sum_{n=1}^{m} \frac{1}{(2n-1)^2} - \frac{1}{2} \sum_{n=1}^{m} \frac{1}{n}$$

$$=: t_m - \frac{1}{2} u_m.$$

The sequence $(t_m)$ converges, by the Comparison Test (compare with the convergent series $\sum (1/n^2)$), so if $(s_{2m})$ converges, then $(u_m)$ converges by the Algebra of Limits. But $u_m$ is the $m^{\text{th}}$ partial sum of the harmonic series, which is known to be divergent. Hence $(s_{2m})$ diverges, so $(s_k)$ diverges. □

Equally, one can write down alternating series for which $(a_n)$ is not decreasing but which do, nevertheless, converge (e.g. modify Example 5.18 by taking $a_n = 1/n^3$ for $n$ even). The point is that the Alternating Series Test is a *sufficient* but *not necessary* test for convergence.

## 5.4 Absolute convergence

One rather crude way to show that some alternating series converge is to throw away the extra information about the sign of the terms altogether.

**Definition 5.19.** A real series $\sum_{n=1}^{\infty} a_n$ **converges absolutely** if $\sum_{n=1}^{\infty} |a_n|$ converges in the usual sense (that is, the sequence of partial sums $s_k = \sum_{n=1}^{k} |a_n|$ converges).

**Remark.** This definition applies to all series, whether alternating or not.

**Theorem 5.20.** *If a real series $\sum_{n=1}^{\infty} a_n$ converges absolutely, then it converges.*

**Proof.** We are given that the series $\sum_{n=1}^{\infty} |a_n|$ converges, and hence its sequence of partial sums is bounded above, by $L$ say. Let

$$b_n = \begin{cases} a_n, & \text{if } a_n \geq 0 \\ 0, & \text{if } a_n < 0 \end{cases} \qquad c_n = \begin{cases} 0, & \text{if } a_n \geq 0 \\ a_n, & \text{if } a_n < 0 \end{cases}$$

and note that $a_n = b_n + c_n$ for all $n$. Let $s_k$, $t_k$, $u_k$ denote the $k^{\text{th}}$ partial sums of the series with terms $a_n, b_n$, and $c_n$, respectively. Then $s_k = t_k + u_k$, and we seek to prove that the sequence $(s_k)$ converges. Now, $(t_k)$ is increasing, since $b_n \geq 0$ for all $n$. Furthermore,

$$t_k = \sum_{n=1}^{k} b_n = \sum_{n=1}^{k} |b_n| \leq \sum_{n=1}^{k} |a_n| \leq L,$$

so, by the Monotone Convergence Theorem, $(t_k)$ converges, to $L_+$ say. Similarly, $(u_k)$ is decreasing, since $c_n \leq 0$ for all $n$, and

$$u_k = \sum_{n=1}^{k} c_n = -\sum_{n=1}^{k} |c_n| \geq -\sum_{n=1}^{k} |a_n| \geq -L$$

so, by the Monotone Convergence Theorem, $(u_k)$ converges, to $L_-$ say. Hence, by Theorem 3.5, $s_k \to L_+ + L_-$. $\qquad \square$

**Remark.** The converse of Theorem 5.20 is false. For example, the alternating harmonic series

$$\sum_{n=1}^{\infty} \frac{(-1)^{n+1}}{n}$$

is convergent but is not absolutely convergent, since in this case

$$\sum_{n=1}^{\infty} \left| \frac{(-1)^{n+1}}{n} \right| = \sum_{n=1}^{\infty} \frac{1}{n},$$

which we have proved is divergent (see Examples 5.2 and 5.17).

**Example 5.21.** Claim: $\sum_{n=1}^{\infty} \dfrac{\sin n}{n^3}$ converges.

*Note that this series is not alternating, so we can't use the Alternating Series Test. Nor are the terms all positive, so we can't use the Comparison or Ratio Tests. What we can do is show that it converges absolutely and use Theorem 5.20*

**Proof.** Let $a_n = n^{-3} \sin n$ and $b_n = 1/n^2$. Then $|a_n| > 0$ and $b_n > 0$ and

$$\frac{|a_n|}{b_n} = \frac{|\sin n|}{n} \le \frac{1}{n} \le 1.$$

Since we know that $\sum_{n=1}^{\infty} b_n$ converges (Example 5.8) it follows that $\sum_{n=1}^{\infty} |a_n|$ converges also, by the Comparison Test, part (i). Hence $\sum_{n=1}^{\infty} a_n$ converges absolutely and so converges by Theorem 5.20. $\square$

**Example 5.22.** Consider the sequence

$$a_n = \begin{cases} \frac{1}{n^3} & n \text{ odd} \\ -\frac{1}{n^2} & n \text{ even} \end{cases}$$

I claim that the series $\sum_{n=1}^{\infty} a_n$ converges.

**Proof.** Let $b_n = 1/n^2$. Then

$$\frac{|a_n|}{b_n} = \begin{cases} \frac{1}{n} & n \text{ odd} \\ 1 & n \text{ even} \end{cases}$$

$$\le 1.$$

Since $|a_n|/b_n$ is bounded above, and $\sum b_n$ is known to converge (Example 5.8, we conclude that $\sum |a_n|$ converges also, by part (i) of the Comparison Test, that is, $\sum a_n$ converges *absolutely*. Hence $\sum a_n$ converges by Theorem 5.20. $\square$

**Remark.** This is an alternating series, but we can't use the Alternating Series Test, because $|a_n|$ is not a decreasing sequence.

## 5.5    Summary

- Given a series $\sum_{n=1}^{\infty} a_n$ we call $a_n$ the $n^{\text{th}}$ **term** of the series, and

$$s_k = \sum_{n=1}^{k} a_n$$

  the $k^{\text{th}}$ **partial sum** of the series.
- A series **converges** to $L$ if and only if its sequence of partial sums $(s_k)$ converges to $L$, in the usual sense: for each $\varepsilon > 0$ there exists $K \in \mathbb{Z}^+$ such that for all $k \geq K$, $|s_k - L| < \varepsilon$. The series **diverges** if $(s_k)$ diverges (that is, does not converge).
- The **harmonic series** $\sum_{n=1}^{\infty} 1/n$ diverges.
- For all $p \geq 2$, the series $\sum_{n=1}^{\infty} 1/n^p$ converges.
- The **geometric series** $\sum_{n=0}^{\infty} \alpha^n$ converges to $1/(1 - \alpha)$ for all $\alpha \in (-1, 1)$.
- The **Divergence Test**: if $\sum_{n=1}^{\infty} a_n$ converges, then $a_n \to 0$.
- The **Comparison Test**: if $a_n, b_n > 0$ and

  (i) $a_n/b_n$ is bounded and $\sum b_n$ converges, then $\sum a_n$ converges.
  (ii) $b_n/a_n$ is bounded and $\sum b_n$ diverges, then $\sum a_n$ diverges.

- The **Ratio Test**: if $a_n > 0$ and $a_{n+1}/a_n \to L$ then

  (i) if $L < 1$, $\sum a_n$ converges.
  (ii) if $L > 1$, $\sum a_n$ diverges.

- The **Alternating Series Test**: if $a_n > 0$, $a_n \to 0$ and $(a_n)$ is decreasing, then $\sum (-1)^{n+1} a_n$ converges.
- A series $\sum a_n$ **converges absolutely** if $\sum |a_n|$ converges.
- Absolute convergence implies convergence. The converse is false.

## 5.6 Tutorial problems

1. Consider the series $\sum_{n=1}^{\infty} a_n$ where $a_n = \dfrac{n+1}{n^2(n+2)^2}$.

   (a) Find a formula for the $k^{\text{th}}$ partial sum, $s_k = \sum_{n=1}^{k} a_n$.

   (b) Hence show that the series converges, and find its limit.

2. Determine whether the following series converge. Rigorously justify your answers (using the convergence tests in sections 5.2 and 5.3).

   (a) $\displaystyle\sum_{n=1}^{\infty} \frac{2^n}{1+3^n}$,
           
   (b) $\displaystyle\sum_{n=1}^{\infty} (-1)^{n+1}\frac{n}{(n+1)(n+2)}$,

   (c) $\displaystyle\sum_{n=1}^{\infty} (-1)^{n+1}\frac{(n+20)^2}{n!}$,
      
   (d) $\displaystyle\sum_{n=1}^{\infty} \frac{8n^2 - 7n + 12}{4n^3 + n^2 + 5n}$.

3. Consider the series $\displaystyle\sum_{n=1}^{\infty} \cos\left(\frac{2n\pi}{3}\right)\frac{1}{3^n}$.

   (a) Prove that the series converges absolutely, and hence converges.

   (b) Prove that its limit is $-\dfrac{5}{26}$.

## 5.7 Homework problems

1. Consider the series $\sum_{n=1}^{\infty} a_n$ where $a_n = \dfrac{1}{n(n+3)}$.

   (a) Find a formula for the $k^{\text{th}}$ partial sum, $s_k = \sum_{n=1}^{k} a_n$. *(Hint: partial fractions!)*

   (b) Hence show that the series converges, and find its limit.

2. Determine whether the following series converge. Rigorously justify your answers (using the convergence tests in sections 5.2 and 5.3).

(a) $\displaystyle\sum_{n=1}^{\infty} \frac{n}{n^2+1}$,  (b) $\displaystyle\sum_{n=1}^{\infty} \frac{2^n}{n!}$,    (c) $\displaystyle\sum_{n=0}^{\infty} \frac{2^n}{1+2^n}$,

(d) $\displaystyle\sum_{n=1}^{\infty} \frac{n^2}{n^4+1}$,  (e) $\displaystyle\sum_{n=1}^{\infty} (-1)^{n+1} \frac{1}{3n+2}$,  (f) $\displaystyle\sum_{n=1}^{\infty} (-1)^{n+1} \frac{|\sin n|}{n^4}$,

(g) $\displaystyle\sum_{n=1}^{\infty} \frac{n!}{(2n)!}$,  (h) $\displaystyle\sum_{n=0}^{\infty} \left( \sum_{m=0}^{n} \frac{1}{2^m} \right)$.

3. Consider the series $\displaystyle\sum_{n=0}^{\infty} \sin\left( \frac{(2n+1)\pi}{4} \right) \frac{1}{2^n}$.

(a) Prove that the series converges absolutely, and hence converges.

(b) Prove that its limit is $\dfrac{3\sqrt{2}}{5}$. *Hint: having shown that the sequence of partial sums* $(s_k)$ *converges, to show that its limit is L it's enough to show that a subsequence of* $(s_k)$ *converges to L (by Theorem 4.3). Consider the subsequence* $(s_{4k-1})$.

# Chapter 6

# Continuous functions

## 6.1 Sequential continuity

Given a function $f : \mathbb{R} \to \mathbb{R}$, what does it mean to say that the function is *continuous*? It's common, when first introduced to the notion of continuity, to be told that a function is continuous "if you can draw its graph without taking your pencil off the paper". Clearly this is *not* an acceptable mathematical definition (what if you have no pencil, or paper, or hands? Does the continuity or otherwise of $f$ depend on your skill as a draughtsman?). So our first job will be to define continuity precisely. In fact, it's quite easy to give a precise (and rather elegant) definition of continuity using nothing more than the notion of convergence of sequences. The definition works equally well when the domain of $f$ is a subset of $\mathbb{R}$, not necessarily the whole of $\mathbb{R}$, so that is how we will formulate it.

**Definition 6.1.** Let $D \subseteq \mathbb{R}$. A function $f : D \to \mathbb{R}$ is **continuous at a point** $a \in D$ if, given any sequence $(x_n)$ in $D$ such that $x_n \to a$, $f(x_n) \to f(a)$. If $f$ is not continuous at $a \in D$, we say that $f$ is **discontinuous** at $a$. The function is **continuous** if it is continuous at $a$ for all $a \in D$.

**Example 6.2.** $f : \mathbb{R} \to \mathbb{R}$ such that $f(x) = c$, constant, is continuous.

**Proof.** Let $a \in \mathbb{R}$ and $x_n \to a$. Then $f(x_n) = c \to c = f(a)$. $\qquad\square$

**Example 6.3.** $f : \mathbb{R} \to \mathbb{R}$ such that $f(x) = x$ is continuous.

**Proof.** Let $a \in \mathbb{R}$ and $x_n \to a$. Then $f(x_n) = x_n \to a = f(a)$. $\qquad\square$

**Example 6.4.** $f : \mathbb{R}\backslash\{0\} \to \mathbb{R}$ such that $f(x) = 1/x$ is continuous.
If you've been taught in the past that a continuous function is "one whose graph you can draw without taking your pen off the paper" then this

example will be surprising at first sight. Nonetheless, this function really is continuous according to our (precise) definition:

**Proof.** Let $a \in \mathbb{R}\backslash\{0\}$ (i.e. let $a$ be a non-zero real number) and $x_n$ be a sequence in $\mathbb{R}\backslash\{0\}$ such that $x_n \to a$. Then $x_n \neq 0$, so

$$f(x_n) = \frac{1}{x_n} \to \frac{1}{a} = f(a)$$

by the Algebra of Limits (Theorem 3.5(iii)).                    $\square$

**Example 6.5.** $f : \mathbb{Z} \to \mathbb{R}$ such that $f(x) = (-1)^x$ is continuous.

This one is also counterintuitive. In fact, it gets worse:

**Example 6.5 (generalized).** *Every* function $f : \mathbb{Z} \to \mathbb{R}$ is continuous!

**Proof.** Let $a \in \mathbb{Z}$ and $(x_n)$ be a sequence in $\mathbb{Z}$ converging to $a$. Then, by the definition of convergence, there exists $N \in \mathbb{Z}^+$ such that, for all $n \geq N$, $|x_n - a| < 1/2$. But $x_n$ and $a$ are both integers, so if they lie distance less than $1/2$ away from one another, they must be equal. Hence, for all $n \geq N$, $x_n = a$. Now let $\varepsilon > 0$ be given. Then for all $n \geq N$, $|f(x_n) - f(a)| = |f(a) - f(a)| = 0 < \varepsilon$. Hence, $f(x_n) \to f(a)$.      $\square$

Don't worry, however, there *are* discontinuous functions:

**Example 6.6.** Let $f : \mathbb{R} \to \mathbb{R}$ such that $f(x) = 0$ if $x < 0$ and $f(x) = 1$ for $x \geq 0$. Then $f$ is discontinuous at 0.

**Proof.** The sequence $x_n = -1/n$ converges to 0 but

$$f(x_n) = f(-1/n) = 0 \to 0 \neq f(0).$$

$\square$

The graph of this function clearly shows the discontinuity at $x = 0$, see Figure 6.1. This is an example of a function which is *discontinuous* at a single isolated point. It's also possible to construct functions which are *continuous* at a single isolated point (and discontinuous everywhere else):

**Example 6.7.** Let $f : \mathbb{R} \to \mathbb{R}$ such that

$$f(x) = \begin{cases} 0, & x \in \mathbb{Q} \\ x, & x \notin \mathbb{Q} \end{cases}.$$

Then $f$ is continuous at 0 and discontinuous everywhere else.

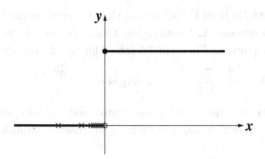

Fig. 6.1 Graph of the step function defined in Example 6.6. The crosses depict a sequence converging to 0 whose image sequence does not converge to $f(0) = 1$. It follows that $f$ is discontinuous at $x = 0$.

**Proof.** Let $a \in \mathbb{R}$ and assume that $f$ is continuous at $a$. By Theorems 1.25 and 1.34, for each $n \in \mathbb{Z}^+$, there exist $r_n \in \mathbb{Q}$ and $i_n \in \mathbb{R}\backslash\mathbb{Q}$ such that $a < r_n < a + 1/n$ and $a < i_n < a + 1/n$. Clearly $r_n \to a$ and $i_n \to a$ by the Squeeze Rule. Hence, by the definition of continuity, $f(r_n) \to f(a)$ and $f(i_n) \to f(a)$. But $r_n$ is rational, so $f(r_n) = 0 \to 0$, and $i_n$ is irrational, so $f(i_n) = i_n \to a$. Hence $a = 0$.

(Aside: so what have we proved? Have we proved that $f$ is continuous at 0? No we have *not!* We have proved that *if $f$ is continuous at $a$ then $a = 0$*. We have *not* proved that *if $a = 0$ then $f$ is continuous at $a$!* Another way to say what we have proved is that *if $a \neq 0$ then $f$ is *not* continuous at $a$*. That is, up to this point, we have proved that $f$ is *discontinuous* at every $a \neq 0$.)

Let $(x_n)$ be any sequence in $\mathbb{R}$ such that $x_n \to 0$. Let $\varepsilon > 0$ be given. Then, since $x_n \to 0$, there exists $N \in \mathbb{Z}^+$ such that for all $n \geq N$, $|x_n - 0| < \varepsilon$. But then, for all $n \geq N$,

$$|f(x_n) - 0| = |f(x_n)| \leq |x_n| < \varepsilon$$

since $f(x_n) = x_n$ if $x_n$ is irrational, and is 0 otherwise. Hence $f(x_n) \to 0 = f(0)$. Hence $f$ is continuous at 0. $\qquad\square$

## 6.2 Basic properties of continuous functions

If we are given a complicated function such as

$$f : \mathbb{R} \to \mathbb{R}, \qquad f(x) = \frac{x^3 - 7x}{x^2 + 1}$$

we can show directly from Definition 6.1 that $f$ is continuous, by considering an arbitrary sequence $(x_n)$ converging to $a$. To do this we will have to appeal at some point to Theorem 3.5 (the Algebra of Limits), to argue that

$$f(x_n) = \frac{x_n^3 - 7x_n}{x_n^2 + 1} \qquad \text{converges to} \qquad \frac{a^3 - 7a}{a^2 + 1} = f(a).$$

The alternative to this is to prove, once and for all, that continuity is "preserved" under sums, products, and (under suitable restrictions) quotients.

**Theorem 6.8 (Algebra Property of Continuous Functions).** *Let* $f : D \to \mathbb{R}$ *and* $g : D \to \mathbb{R}$ *be continuous at* $a \in D$. *Then*

*(i)* $f + g$ *is continuous at* $a$.
*(ii)* $fg$ *is continuous at* $a$.

*If, in addition,* $f(x) \neq 0$ *for all* $x \in D$, *then*

*(iii)* $1/f$ *is continuous at* $a$.

**Proof.** Given any sequence $(x_n)$ in $D$ such that $x_n \to a$, we know that $f(x_n) \to f(a)$ and $g(x_n) \to g(a)$ (since $f, g$ are continuous at $a$). Hence

$$(f + g)(x_n) = f(x_n) + g(x_n) \to f(a) + g(a) \qquad \text{(by Theorem 3.5(i)),}$$
$$(fg)(x_n) = f(x_n)g(x_n) \to f(a)g(a) \qquad \text{(by Theorem 3.5(ii)),}$$
$$\text{and} \qquad (1/f)(x_n) = \frac{1}{f(x_n)} \to \frac{1}{f(a)} \qquad \text{(by Theorem 3.5(iii))}$$

where, in the final line, we have assumed $f(x_n) \neq 0$ and $f(a) \neq 0$. Hence $f + g$, $fg$, and $1/f$ are continuous at $a$. $\qquad \square$

Recall that a **polynomial** function is a function of the form

$$p(x) = a_0 + a_1 x + a_2 x^2 + \cdots + a_m x^m$$

where $a_0, \ldots, a_m$ are (real) constants. If $a_m \neq 0$, we say that $p$ has *degree* $m$. Polynomials are a very nice class of functions. In particular, they are always continuous:

**Proposition 6.9.** *Every polynomial function* $p : \mathbb{R} \to \mathbb{R}$ *is continuous.*

**Proof.** We prove this by induction on $m$, the degree of the polynomial $p$. So, if $m = 0$, then $p(x)$ is a degree 0 polynomial, that is, $p(x) = a_0$, constant. Hence $p(x)$ is continuous (see Example 6.2). Now assume that every polynomial of degree $k$ is continuous. Let $p(x)$ be a degree $k + 1$

polynomial. Then $p(x) = xq(x) + a_0$ where $q(x)$ is a polynomial of degree $k$, and so is, by assumption, continuous. Now $f(x) = x$ is continuous (see Example 6.3), so $p(x) = xq(x) + a_0$ is continuous by Theorem 6.8(i),(ii). Hence, by induction, every polynomial is continuous □

**Example 6.10 (Rational functions).** Let $p(x)$, $q(x)$ be polynomials, and define $D = \{x \in \mathbb{R} : q(x) \neq 0\}$. Then the function

$$f : D \to \mathbb{R}, \qquad f(x) = \frac{p(x)}{q(x)}$$

is said to be **rational** (it is a quotient of polynomials). By Proposition 6.9 and Theorem 6.8(iii), every rational function is continuous. We've already seen an example of this:

$$f : \mathbb{R}\backslash\{0\} \to \mathbb{R}, \qquad f(x) = \frac{1}{x}$$

is continuous. □

A nice thing about the set of polynomials is that it is *closed* under function composition. That is, if $p$ and $q$ are polynomials, so is $q \circ p$ (recall that $(q \circ p)(x) = q(p(x))$). For example:

$$p(x) = x^3 + 1, \qquad q(x) = 3x^2 - 2$$
$$\Rightarrow (q \circ p)(x) = q(x^3 + 1) = q(x^3 + 1)$$
$$= 3(x^6 + 2x^3 + 1) - 2 = 3x^6 + 6x^3 + 1.$$

So, given a pair of polynomials (certainly continuous), their composition is a polynomial (and hence is continuous). Actually, rational functions are also closed under composition (though we have to be careful about the domain of the composite functions to make this precise). So, given a pair of rational functions (certainly continuous), their composition is rational (and hence is continuous). Is this a special feature of polynomials and rational functions? No: it follows almost immediately from our definition of continuity that the composition of two continuous functions is continuous.

**Theorem 6.11.** *Let $D, E \subseteq \mathbb{R}$, $f : D \to E$ and $g : E \to \mathbb{R}$. If $f$ is continuous at $a \in D$ and $g$ is continuous at $f(a) \in E$, then $g \circ f$ is continuous at $a$.*

**Proof.** Let $(x_n)$ be any sequence in $D$ which converges to $a$. Then $f(x_n) \to f(a)$, since $f$ is continuous at $a$, so

$$(g \circ f)(x_n) = g(f(x_n)) \to g(f(a)) = (g \circ f)(a),$$

since $g$ is continuous at $f(a)$. Hence $g \circ f$ is continuous at $a$. □

At the moment, we don't have many functions which we know (rigorously) are continuous (polynomials and rational functions are pretty much it). We will see shortly that the functions

$$f : [0, \infty) \to \mathbb{R}, \qquad f(x) = x^{1/p}$$

are well-defined and continuous (for any positive integer $p$). Once we know this, it will follow immediately from Theorem 6.11 that a complicated beast like

$$f : [0, \infty) \to \mathbb{R}, \qquad f(x) = \frac{\sqrt{x + \sqrt{x + 1}}}{1 + x^{1/3} + x^{5/6}}$$

is continuous. Let's ask ourselves, for a given $x \geq 0$, what does $x^{1/p}$ actually mean? It is, by definition, that (non-negative) real number whose $p^{\text{th}}$ power is $x$, that is, $x^{1/p} = y$ if $y \geq 0$ and $y^p = x$. To prove that such a number $y$ exists we will appeal to a big theorem: the Intermediate Value Theorem.

## 6.3   The Intermediate Value Theorem

The following theorem is truly fundamental. It is the basis of a useful method for approximately solving equations (the bisection method) and, as we will see, implies the existence of square roots and other similar functions.

**Theorem 6.12 (Intermediate Value Theorem).** *Let $f : [a, b] \to \mathbb{R}$ be continuous and $y$ be any real number between $f(a)$ and $f(b)$. Then there exists $c \in [a, b]$ such that $f(c) = y$.*

**Proof.** There are three possibilities: (i) $f(a) = f(b)$, (ii) $f(a) < f(b)$, and (iii) $f(a) > f(b)$.

(i) If $f(a) = f(b)$ then the only real number between them is $y = f(a)$, so $c = a$ will do.

(ii) Assume $f(a) < f(b)$. If $f(b) = y$ then $c = b$ will do. So assume $y < f(b)$. Define the set

$$A = \{x \in [a, b] : f(x) \leq y\},$$

which is nonempty (since $a \in A$) and bounded above (by $b$). Hence, by the Axiom of Completeness (Axiom 1.22), there exists $c$, the supremum of $A$. We will prove that $c \in [a, b]$ and that $f(c) = y$, which completes case (ii). $c$ is an upper bound on $A$ and $a \in A$, so $a \leq c$. Further, $b$ is also an upper bound on $A$, and $c$ is the *least* upper bound on $A$, so $c \leq b$. Hence $c \in [a, b]$. Consider $c - 1/n$, where $n \in \mathbb{Z}^+$. Since this is less than $c$, it is *not*

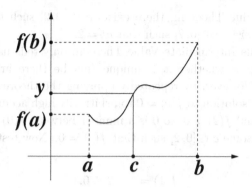

Fig. 6.2   Graphical representation of the Intermediate Value Theorem.

an upper bound on $A$, so there exists $x_n \in A$ with $x_n > c - 1/n$. But then $c - 1/n \le x_n \le c$, so $x_n \to c$ by the Squeeze Rule. $f$ is continuous at $c$, so $f(x_n) \to f(c)$. But each $x_n \in A$, so $f(x_n) \le y$ for all $n$, and hence its limit cannot exceed $y$, by Proposition 3.4. Hence $f(c) \le y$. Now $f(b) > y$, so $c \ne b$, and hence $c < b$. Consider now the sequence $z_n = c + (b-c)/n$. This is a sequence in $[a, b]$, each term of which exceeds $c$, so for all $n$, $z_n \notin A$ (recall $c$ is an upper bound on $A$). Hence $f(z_n) \ge y$ for all $n$. But $z_n \to c$ and $f$ is continuous, so $f(z_n) \to f(c)$. Hence, by Proposition 3.4, $f(c) \ge y$. Since we've already shown that $f(c) \le y$, it follows that $f(c) = y$.

(iii) If $f(a) > f(b)$, define $\tilde{f} : [a, b] \to \mathbb{R}$ such that $\tilde{f}(x) = -f(x)$ and $\tilde{y} = -y$. Then $\tilde{f}$ is continuous, $\tilde{f}(a) < \tilde{f}(b)$, and $\tilde{y}$ is between $\tilde{f}(a)$ and $\tilde{f}(b)$, so, as proved in part (ii), there exists $c \in [a, b]$ such that $\tilde{f}(c) = \tilde{y}$. But then $f(c) = y$. $\qquad\square$

A graphical representation of the statement of the Intermediate Value Theorem is given in Figure 6.2. Note that the assumption that the domain of $f$ is an interval is crucial:

**Example 6.13.** $f : \mathbb{R} \setminus \{0\} \to \mathbb{R}$, $f(x) = 1/x$ is continuous (by Example 6.4, or Example 6.10). Let $a = -1$ and $b = 1$. Then $f(a) = -1$ and $f(b) = 1$ and so 0 is a number between $f(a)$ and $f(b)$. But there is no $c$ such that $f(c) = 0$ ($1/c = 0$ has no solution). This is not a counterexample to the Intermediate Value Theorem, since $f$ is *not* continuous on $[-1, 1]$. $\qquad\square$

**Example 6.14.** Let $f : \mathbb{R} \to \mathbb{R}$ such that $f(x) = x^5 - 2$. Then $f$ is continuous. In particular, it is continuous on $[0, 2]$. Now $f(0) = -2$ and $f(2) = 30$, so 0 is a number between $f(0)$ and $f(2)$. Hence, by the

Intermediate Value Theorem, there exists $c \in [0, 2]$ such that $f(c) = 0$. That is, there exists $c \in [0, 2]$ such that $c^5 = 2$.

Note that the Intermediate Value Theorem only tells us that $c$ exists, not what it is, or whether it is unique (maybe there are hundreds of different $c$'s!). However, by repeatedly applying the theorem, we can find an approximate solution to $f(c) = 0$ to arbitrarily high accuracy. We know that $f(0) < 0$ and $f(2) > 0$, so $0$ is a number between $f(0)$ and $f(2)$, and hence, there is some $c \in [0, 2]$ such that $f(c) = 0$. Now test the midpoint of $[0, 2]$:

$$f(1) = 1 - 2 < 0.$$

So $0$ is a number between $f(1)$ and $f(2)$, and $f$ is continuous on $[1, 2]$, so by the Intermediate Value Theorem, there exists $c \in [1, 2]$ such that $f(c) = 0$. (Had $f(1)$ been positive, we'd have deduced that $c \in [0, 1]$. Had it been $0$, we'd have deduced that $c = 1$ exactly!) Now test the midpoint of $[1, 2]$, and so on:

$$f(1.5) = 5.59375 > 0 \quad \Rightarrow \quad c \in [1, 1.5],$$
$$f(1.25) = 1.051757812 > 0 \quad \Rightarrow \quad c \in [1, 1.25],$$
$$f(1.125) = -0.197967529 < 0 \quad \Rightarrow \quad c \in [1.125, 1.25].$$

This is called the "bisection method". Each time we test the midpoint and apply the Intermediate Value Theorem, we halve the width of the interval in which we've "trapped" our solution $c$. We initially know $c$ is in $[0, 2]$ an interval of width $2$, so after $n$ iterations of the method, we know that $c$ is in some interval of width $2 \times (1/2)^n = 1/2^{n-1}$ (check: after $n = 4$ iterations, we should have $c$ trapped in an interval of width $1/8$ – correct!). Let's say we need to know $c$ to within an error of $5 \times 10^{-5}$. Then we need to trap it in an interval of width less than $0.00005$. How many iterations of the method do we need? The answer is any $n$ for which

$$\frac{1}{2^{n-1}} < 0.00005$$

will do. We know that such an $n$ exists because the sequence $1/2^{n-1}$ converges to $0$. If we allow ourselves to use logarithms, we can solve the above inequality. Otherwise, trial and error (substituting values of $n$) shows that $n = 16$ is the smallest positive integer satisfying the inequality. So, not only will the bisection method give us the (approximate) solution to the accuracy we require, we can figure out before we start how many iterations we'll need to make to get this accuracy. $\quad\square$

By the way, we've just proved that 2 has a fifth root! Our next endeavour will be to generalize this argument so as to prove the existence of square, and other, roots of any non-negative number. For this, we will need the following simple definition and fact.

**Definition 6.15.** Let $D \subseteq \mathbb{R}$. A function $f : D \to \mathbb{R}$ is **strictly increasing** if, for all $x, y \in D$, if $x < y$ then $f(x) < f(y)$. Similarly, $f$ is **strictly decreasing** if, for all $x, y \in D$, if $x < y$ then $f(x) > f(y)$.

An immediate consequence of this definition is that strictly increasing and strictly decreasing functions are injective.

**Lemma 6.16.** *Let $f : D \to \mathbb{R}$ be strictly increasing or strictly decreasing. Then $f$ is injective.*

**Proof.** Assume $f$ is strictly increasing and let $x, y$ be any elements of $D$ such that $f(x) = f(y)$. If $x < y$ then $f(x) < f(y)$, a contradiction, and if $y < x$ then $f(y) < f(x)$, a contradiction. Hence $x = y$, so $f$ is injective. The case of strictly decreasing $f$ is handled similarly. $\square$

**Lemma 6.17.** *Let $p \in \mathbb{Z}^+$ and $f : [0, \infty) \to \mathbb{R}$ such that $f(x) = x^p$. Then $f$ is strictly increasing.*

**Proof.** Let $x, y$ be any real numbers such that $0 \le x < y$. Then
$$f(x) - f(y) = x^p - y^p = (x - y)(x^{p-1} + x^{p-2}y + \cdots + xy^{p-2} + y^{p-1}).$$
Now $x - y < 0$, and
$$x^{p-1} + x^{p-2}y + \cdots + xy^{p-2} + y^{p-1} \ge y^{p-1} > 0,$$
since $x \ge 0$ and $y > 0$, so $f(x) - f(y) < 0$. Hence $f(x) < f(y)$. $\square$

**Proposition 6.18.** *Let $p \in \mathbb{Z}^+$. Given any $y \ge 0$, there exists a unique $x \ge 0$ such that $x^p = y$. We denote this real number $y^{1/p}$ (or, if $p = 2$, $\sqrt{y}$) and call it the $p^{th}$ root of $y$.*

**Proof.** Consider the function $f : [0, 1 + y] \to \mathbb{R}$, $f(x) = x^p$. This, being a polynomial, is continuous. Now $f(0) = 0 \le y$. Further, I claim that $f(1+y) > y$ since, if $y \ge 1$ then $f(1+y) > f(y) = y^p \ge y$ (by Lemma 6.17), while if $y < 1$ then $f(1 + y) \ge f(1) = 1 > y$ (again by Lemma 6.17). So $y$ is a real number between $f(0)$ and $f(1 + y)$. Hence, by the Intermediate Value Theorem, there exists $x \in [0, 1 + y]$ such that $f(x) = y$.

Let $z \in [0, \infty)$ also satisfy $f(z) = y$. Then $f(z) = f(x)$. But $f$ is strictly increasing, hence injective (Lemma 6.16), so $z = x$. Hence, the real number in $[0, \infty)$ whose $p^{th}$ power is $y$ is unique. $\square$

We've now proved that there exists a function

$$g : [0, \infty) \to \mathbb{R}, \qquad g(x) = x^{1/p}$$

for any positive integer $p$. You won't be surprised to learn that this function is, like $x^p$, strictly increasing on $[0, \infty)$.

**Lemma 6.19.** *Let* $g : [0, \infty) \to \mathbb{R}$ *such that* $g(x) = x^{1/p}$. *Then* $g$ *is strictly increasing.*

**Proof.** Assume, to the contrary, that $0 \leq x < y$ but $g(x) \geq g(y)$. Let $f(x) = x^p$. Then, by Lemma 6.17,

$$f(g(y)) \leq f(g(x)).$$

But $f(g(y)) = y$ and $f(g(x)) = x$, so $y \leq x$, a contradiction. $\qquad\square$

You will also not be surprised to learn that $g$ is continuous. Proving this is slightly tricky. We'll do the case $p = 2$ first, since this illustrates the required trick, then return to the general case.

**Proposition 6.20.** *The function* $g : [0, \infty) \to \mathbb{R}$, $g(x) = \sqrt{x}$ *is continuous.*

**Proof.** Let $a \in (0, \infty)$ (we will handle the case $a = 0$ separately). We must show that, given any sequence $x_n \in [0, \infty)$ converging to $a$, $g(x_n)$ converges to $g(a)$. So, let $(x_n)$ be such a sequence, and let $\varepsilon > 0$ be given. Then there exists $N \in \mathbb{Z}^+$ such that, for all $n \geq N$, $|x_n - a| < \varepsilon\sqrt{a}$. Now for all $n \geq N$,

$$|g(x_n) - g(a)| = |\sqrt{x_n} - \sqrt{a}| = \left| \frac{(\sqrt{x_n} - \sqrt{a})(\sqrt{x_n} + \sqrt{a})}{\sqrt{x_n} + \sqrt{a}} \right|$$

$$= \frac{|x_n - a|}{\sqrt{x_n} + \sqrt{a}} \leq \frac{1}{\sqrt{a}}|x_n - a| < \varepsilon.$$

Hence $g(x_n) \to g(a)$.

Finally, let $(x_n)$ be a sequence in $[0, \infty)$ such that $x_n \to 0$. We must show that $g(x_n) \to g(0) = 0$. Let $\varepsilon > 0$ be given. Then there exists $N \in \mathbb{Z}^+$ such that, for all $n \geq N$, $0 \leq x_n < \varepsilon^2$. Now for all $n \geq N$,

$$|g(x_n) - g(0)| = |\sqrt{x_n}| = \sqrt{x_n}$$
$$< \sqrt{\varepsilon^2} \qquad \text{by Lemma 6.19}$$
$$= \varepsilon.$$

Hence $g(x_n) \to g(0)$. $\qquad\square$

The trick here was to rewrite $\sqrt{x_n} - \sqrt{a}$ using the factorization formula for a difference of two squares. In the general $p$ case, the analogue of the difference of two squares is the formula

$$x^p - y^p = (x - y)(x^{p-1} + x^{p-2}y + \cdots + xy^{p-2} + y^{p-1})$$

which we already used to prove Lemma 6.17.

**Proposition 6.21.** *The function* $g : [0, \infty) \to \mathbb{R}$, $g(x) = x^{1/p}$ *is continuous.*

**Proof.** Let $a \in (0, \infty)$ (we will handle the case $a = 0$ separately). We must show that, given any sequence $x_n \in [0, \infty)$ converging to $a$, $g(x_n)$ converges to $g(a)$. So, let $(x_n)$ be such a sequence, and let $\varepsilon > 0$ be given. Let $y_n = g(x_n) = x_n^{1/p}$ and $b = g(a) = a^{1/p}$. Then

$$y_n^p - b^p = (y_n - b)(y_n^{p-1} + y_n^{p-2}b + \cdots + y_n b^{p-2} + b^{p-1})$$

$$\Rightarrow \quad y_n - b = \frac{y_n^p - b^p}{y_n^{p-1} + y_n^{p-2}b + \cdots + y_n b^{p-2} + b^{p-1}}$$

$$\Rightarrow \quad |y_n - b| \leq \frac{|y_n^p - b^p|}{b^{p-1}}$$

$$\Rightarrow \quad |g(x_n) - g(a)| \leq \frac{|x_n - a|}{b^{p-1}}.$$

Since $x_n \to a$, there exists $N \in \mathbb{Z}^+$ such that, for all $n \geq N$, $|x_n - a| < \varepsilon b^{p-1}$. Now for all $n \geq N$,

$$|g(x_n) - g(a)| \leq \frac{|x_n - a|}{b^{p-1}} < \varepsilon.$$

Hence $g(x_n) \to g(a)$.

Finally, let $(x_n)$ be a sequence in $[0, \infty)$ such that $x_n \to 0$. We must show that $g(x_n) \to g(0) = 0$. Let $\varepsilon > 0$ be given. Then there exists $N \in \mathbb{Z}^+$ such that, for all $n \geq N$, $|x_n - 0| < \varepsilon^p$. Now for all $n \geq N$,

$$|g(x_n) - g(0)| = g(x_n) < g(\varepsilon^p) = \varepsilon$$

by Lemma 6.19. Hence $g(x_n) \to g(0)$. □

Of course, having defined $x^{1/p}$, we can now define $x^r$ for any rational number $r = p/q$:

$$x^{p/q} = (x^{1/q})^p.$$

Strictly speaking, we should check that this definition doesn't depend on which integers $p, q$ we choose to "represent" $r = p/q$. For example, we should check that $x^{2/6}$ is the same as $x^{1/3}$. While we're at it, we should also probably check that $(x^{1/q})^p$ is the same as $(x^p)^{1/q}$ and that $x^{r+r'} = x^r x^{r'}$ and $x^{rr'} = (x^r)^{r'}$ etc. Having established that $x^{1/p}$ *exists*, and is *unique*, verifying all these facts is an exercise in algebra which I invite the interested reader to attempt. We have bigger fish to fry.

## 6.4   The Extreme Value Theorem

In this section we will prove another very important property of continuous functions on closed bounded intervals: they are bounded (above and below) and, furthermore, they attain both a maximum and a minimum value.

**Definition 6.22.** Let $D \subseteq \mathbb{R}$ and $f : D \to \mathbb{R}$. We say that $f$ is **bounded above** if its range $f(D) \subseteq \mathbb{R}$ is bounded above, that is, if there exists $K \in \mathbb{R}$ such that, for all $x \in D$, $f(x) \leq K$. In this case, we define the **supremum** of $f$, $\sup f$, to be the supremum of its range $f(D)$ (that is, the least upper bound on the set $f(D)$). Similarly, we say that $f$ is **bounded below** if its range $f(D) \subseteq \mathbb{R}$ is bounded below, that is, if there exists $L \in \mathbb{R}$ such that, for all $x \in D$, $f(x) \geq L$. In this case, we define the **infimum** of $f$, $\inf f$, to be the infimum of its range $f(D)$ (that is, the greatest lower bound on the set $f(D)$). The function is **bounded** if it is bounded above and below.

   If there exists $a \in D$ such that $f(x) \leq f(a) = M$ for all $x \in D$, we say that $f$ **attains a maximum value of $M$ at** $a$. Similarly, if there exists $b \in D$ such that $f(x) \geq f(b) = L$, we say that $f$ **attains a minimum value of $L$ at** $b$.                                                  □

**Remark.** If a function attains a maximum value, $M$, say, then it is bounded above, since $f(x) \leq M$ for all $x \in D$. Hence, by the Axiom of Completeness, $f$ has a supremum. In fact, $\sup f = M$. (Why? Since $f$ attains a maximum of $M$, there exists $a \in D$ such that $f(a) = M$. Hence, every $K < M$ is less than $f(a)$, so is not an upper bound on $f$. Hence, $M$ is the *least* upper bound on $f$.) Note, however, that *not every function which is bounded above attains a maximum value!*

**Example 6.23.** Let $f : (0, \infty) \to \mathbb{R}$ such that $f(x) = -1/x$. Then $f$ is bounded above, by 0 for example (since $x > 0$, $-1/x < 0$). But it does not attain a maximum value because, given any $a \in (0, \infty)$, $f(a + 1) = -1/(a+1) > -1/a = f(a)$. Of course, this function has a *supremum*: it must do, by the Axiom of Completeness. We've seen that 0 is an upper bound, and given any $K < 0$, $-2/K \in (0, \infty)$ and

$$f(-2/K) = \frac{-1}{(-2/K)} = \frac{K}{2} > K \qquad \text{(careful - remember } K < 0!)$$

so $K$ is not an upper bound on $f$. Hence 0 is the supremum of $f$. But I emphasize again: 0 is *not* the maximum of $f$ and, in fact, $f$ has no maximum value!

Clearly $f$ has no minimum value either, since it isn't even bounded below. Let's prove it: assume to the contrary that $K$ is a lower bound on $f$. Since $f(1) = -1$, $K \leq -1$. But then $x = -1/(2K) \in (0, \infty)$ and $f(x) = 2K < K$, a contradiction. $\square$

The function above is continuous, but bad things happen at the "edges" of its domain (as $x$ gets big, $f(x)$ gets close to 0 but never reaches it, and as $x$ gets close to 0, $f(x)$ grows unbounded below). We now prove that such bad things cannot happen for a continuous function on a closed, bounded interval.

**Theorem 6.24.** *Let $f : [a, b] \to \mathbb{R}$ be continuous. Then $f$ is bounded above. Furthermore, there exists $c \in [a, b]$ such that $f(c) = \sup f$.*

**Proof.** Assume, to the contrary, that $f$ is unbounded above. Then, for each $n \in \mathbb{Z}^+$, $n$ is *not* an upper bound on $f$, so there exists $x_n \in [a, b]$ such that $f(x_n) > n$. The sequence $(x_n)$ is bounded, and hence, by the Bolzano–Weierstrass Theorem, has a convergent subsequence, $(x_{n_k})$, converging to $x$, say. Since $a \leq x_{n_k} \leq b$ for all $k$, we know that $x \in [a, b]$, by Proposition 3.4. Now $f$ is continuous, so $f(x_{n_k}) \to f(x)$, and hence $f(x_{n_k})$ is bounded (Proposition 3.3). But, by definition $f(x_{n_k}) > n_k \geq k$, so $f(x_{n_k})$ is unbounded, a contradiction. Hence, $f$ is bounded above.

Since $A = \{f(t) : t \in [a, b]\}$ is nonempty and bounded above, it has a supremum, $L$ say, by the Axiom of Completeness. It remains to show that there exists $c \in [a, b]$ such that $f(c) = L$. Given any $n \in \mathbb{Z}^+$, $L - 1/n < L$ so is *not* an upper bound on $A$. Hence, there exists $y_n \in [a, b]$ such that $L - 1/n < f(y_n) \leq L$. Since $y_n$ is bounded, it has a convergent subsequence $y_{n_k}$ (by the Bolzano-Weierstrass Theorem). Call its limit $c$. By Proposition 3.4, $c \in [a, b]$. Further $L - 1/n_k < f(y_{n_k}) \leq L$, so $f(y_{n_k}) \to L$ by the Squeeze Rule. But $f$ is continuous, so $f(y_{n_k}) \to f(c)$. Hence $f(c) = L$. $\square$

**Corollary 6.25 (Extreme Value Theorem).** *Let $f : [a, b] \to \mathbb{R}$ be continuous. Then $f$ is bounded and attains both a maximum and a minimum value.*

**Proof.** By Theorem 6.24, $f$ is bounded above, and there exists $c \in [a, b]$ such that $f(c) = M$ where $M$ is the supremum of $f$. By definition (of supremum), $f(x) \leq f(c)$ for all $x \in [a, b]$, so $f$ attains a maximum value (of $M$ at $c$).

The function $g : [a,b] \to \mathbb{R}$, $g(x) = -f(x)$ is continuous, so by Theorem 6.24, is also bounded above and there exists $d \in [a,b]$ such that $g(d) = L$, where $L$ is the supremum of $g$. By definition (of supremum), $g(x) \leq g(d)$ for all $x \in [a,b]$. Hence, $f(x) \geq f(d)$ for all $x \in [a,b]$, so $f$ attains a minimum value (of $-L$ at $d$). $\qquad\square$

**Example 6.26.** The rational function

$$f : [-1,1] \to \mathbb{R}, \qquad f(x) = \frac{1}{x^2 + 1}$$

is continuous (see Example 6.10) and hence, by the Extreme Value Theorem, is bounded and attains both a maximum and a minimum value. In this case, it's not hard to show this directly: for all $x \in [-1,1]$, $x^2 + 1 \geq 1$, so $f(x) \leq f(0) = 1$, and hence $f$ attains a maximum value of 1 at 0. Similarly, for all $x \in [-1,1]$, $1 + x^2 \leq 2$, so $f(x) \geq f(1) = 1/2$, and hence $f$ attains a minimum value of $1/2$ at 1. $\qquad\square$

**Example 6.27.** The rational function

$$f : [-13,-1] \to \mathbb{R}, \qquad f(x) = \frac{x^{27} - 99x^{20} + 45x^{11} - 14x^7 - 5}{x^{12}(x-1)^{15}(x^{80} + 1)^{99}}$$

is continuous (see Example 6.10) and hence, by the Extreme Value Theorem, is bounded and attains both a maximum and a minimum value. Showing this directly would be quite fiddly. $\qquad\square$

## 6.5 Summary

- A function $f : D \to \mathbb{R}$ is **continuous** at $a \in D$ if, for all sequences $(x_n)$ in $D$ converging to $a$, $f(x_n)$ converges to $f(a)$.
- A function $f : D \to \mathbb{R}$ is **continuous** if it is continuous at every point $a \in D$.
- Sums, products, quotients, and compositions of continuous functions are continuous. It follows that all polynomials and rational functions (quotients of polynomials) are continuous.
- The **Intermediate Value Theorem**: if $f : [a, b] \to \mathbb{R}$ is continuous and $y \in \mathbb{R}$ is a number between $f(a)$ and $f(b)$ then there exists $c \in [a, b]$ such that $f(c) = y$.
- We can use the Intermediate Value Theorem to show that square (and other) roots exist. The function $f(x) = x^{1/p}$ is continuous on $[0, \infty)$.
- The **Extreme Value Theorem**: if $f : [a, b] \to \mathbb{R}$ is continuous, then $f$ is bounded and attains both a maximum and a minimum value.

## 6.6  Tutorial problems

1. Given a function $f : \mathbb{R} \to \mathbb{R}$ and a number $a \in \mathbb{R}$, we have a uniquely defined sequence obtained by *iterating* $f$ on the starting value $a$, that is, the sequence $(x_n)$ where

$$x_0 = a, \qquad x_n = f(x_{n-1}) \quad \text{for all } n \geq 1.$$

   (a) Show that if $f$ is continuous and $(x_n)$ converges, its limit, $L$ say, is a **fixed point** of $f$ (that is, $f(L) = L$).

   (b) Consider the case $f(x) = x^2$. Determine the set of values of $a$ for which the sequence $(x_n)$ converges.

   (c) Consider the case $f(x) = x^2 + c$, where $c$ is a real constant. Show that the sequence $(x_n)$ generated by this function never (that is, for no choice of $a$) converges if $c > 1/4$.

   (d) Write down a function $f : \mathbb{R} \to \mathbb{R}$ and a number $a$ such that $(x_n)$ converges but its limit is *not* a fixed point of $f$. (*Hint: your $f$ must be discontinuous, by part (a)!*)

2. Let $f : \mathbb{R} \to \mathbb{R}$ such that $f(x) = x^3 - x^2 - 1$.

   (a) Use the Intermediate Value Theorem to show that $f$ has a root between 1 and 2.

   (b) Use the bisection method to find this root to one decimal place.

   (c) Assume you require the root to within an accuracy of $5 \times 10^{-7}$. How many iterations of the bisection method would be required?

## 6.7  Homework problems

1. Let $f : D \to D$ (so $f$ maps a set to itself). An element $x \in D$ is called a **fixed point** of $f$ if $f(x) = x$.

   (a) Find all fixed points of the map $f : \mathbb{R} \to \mathbb{R}$, $f(x) = x^3 + x^2$.

   (b) Let $f : [0, 1] \to [0, 1]$ be continuous. Prove that $f$ has a fixed point. (*Hint: use the Intermediate Value Theorem – but not for $f$!*)

   (c) Write down a *discontinuous* map $f : [0, 1] \to [0, 1]$ which has no fixed points.

   (d) Write down a *continuous* map $f : [0, 1) \to [0, 1)$ which has no fixed points.

2. Let $f : \mathbb{R} \to \mathbb{R}$ such that $f(x) = x^5 - x^3 + 2x^2 + 1$.

   (a) Use the Intermediate Value Theorem to show that $f$ has a root between $-2$ and $-1$.

(b) Use the bisection method to find this root to one decimal place.

(c) Assume you require the root to within an accuracy of $5 \times 10^{-9}$. How many iterations of the bisection method would be required?

3. Let $f : \mathbb{R} \to \mathbb{R}$ such that $f(x) = -x$ if $x$ is rational and $f(x) = x + 2$ if $x$ is irrational. Prove that $f$ is continuous at $a$ if and only if $a = -1$. (*Hint: this question is very similar to Example 6.7.*)

# Chapter 7

# Some symbolic logic

## 7.1 Statements and their symbolic manipulation

In this book we have dealt with many mathematical statements, some of which were quite logically subtle. For example, consider the statement

"The real sequence $(a_n)$ converges to $L$."

We learned that this statement really means

"For each positive real number $\varepsilon$ there exists a positive integer $N$ such that, for all $n \geq N$, $|a_n - L| < \varepsilon$."

What does the statement

"The real sequence $(a_n)$ converges"

mean? It means that $(a_n)$ converges to $L$ for some real number $L$. That is,

"There exists a real number $L$ such that, for each positive number $\varepsilon$, there exists a positive integer $N$ such that, for all $n \geq N$, $|a_n - L| < \varepsilon$."

Recall that a sequence *diverges* if it doesn't converge. So what does the statement

"The real sequence $(a_n)$ diverges"

really mean? We will see that it means precisely

"For each real number $L$ there exists a positive real number $\varepsilon$ such that, for all positive integers $N$, there exists $n \geq N$ such that $|a_n - L| \geq \varepsilon$."

This is, to put it mildly, *not* obvious! In this chapter we will introduce some basic ideas from symbolic logic which will allow us to manipulate and analyze complicated mathematical statements like "$(a_n)$ converges", and obtain from them related statements like "$(a_n)$ diverges".

The key idea is to represent mathematical statements by symbols which can be combined and algebraically manipulated. For our purposes, a

**statement** is any declaration which is unambiguously either true or false. For example, the declarations

$P$:   $(-2)^2 = 4$
$Q$:   South Korea is smaller in area than the island of Cuba
$R$:   $-2 > 2$
$S$:   Every even integer greater than 2 can be expressed
as the sum of two primes

are all statements: $P$ and $Q$ are true, $R$ is false, and no-one knows (at the time of writing) whether $S$ is true or false, but it is clearly one or the other.[1] On the other hand, declarations like

$P'$:   Tough on crime, tough on the causes of crime!
$Q'$:   Lucy in the sky with diamonds

are not statements because neither can be meaningfully said to be true or false.

Given one or more statements $P, Q, \ldots$, we can produce new statements from them by three basic constructions:

- **Negation**: not $P$, denoted $\neg P$, is the statement which is true if and only if $P$ is false. For the examples above, we have

  $\neg P$:   $(-2)^2 \neq 4$
  $\neg Q$:   South Korea is not smaller in area than the island of Cuba
  $\neg R$    $-2 \leq 2$.

  Note that $\neg P$ and $\neg Q$ are false, while $\neg R$ is true.
- **Conjunction**: $P$ and $Q$, denoted $P \wedge Q$, is the statement which is true if both $P$ and $Q$ are true, and false otherwise. So, for the examples above, $P \wedge Q$ is true, but $P \wedge R$ is false (because $R$ is false).
- **Disjunction**: $P$ or $Q$, denoted $P \vee Q$, is the statement which is true if $P$ is true, or $Q$ is true, or both, but is false if both $P$ and $Q$ are false. So, for the examples above, $P \vee Q$ is true, as is $P \vee R$ (because $P$ is true). However, $(\neg P) \vee R$ is false (because both $\neg P$ and $R$ are false).

You should bear in mind that in mathematics we use the word "or" in its non-exclusive sense. That is "$P$ or $Q$" means "$P$ or $Q$ (or both)", which is different from its use in everyday English. For example, a child who is told

"You will finish your broccoli or you will get no ice cream"

---

[1]This is the infamous Goldbach Conjecture.

could reasonably expect to get some ice cream if he/she finishes his/her broccoli. If the nagging adult is a mathematician, the child would be wise to seek clarification.

It is useful to summarize these definitions using **truth tables**. These represent the truth/falsity of a constructed statement in terms of the truth/falsity of its constituent pieces. We represent "false" by the value 0 and "true" by the value 1:

| $P$ | $\neg P$ |
|---|---|
| 0 | 1 |
| 1 | 0 |

| $P$ | $Q$ | $P \wedge Q$ |
|---|---|---|
| 0 | 0 | 0 |
| 0 | 1 | 0 |
| 1 | 0 | 0 |
| 1 | 1 | 1 |

| $P$ | $Q$ | $P \vee Q$ |
|---|---|---|
| 0 | 0 | 0 |
| 0 | 1 | 1 |
| 1 | 0 | 1 |
| 1 | 1 | 1 |

Two statements $P, Q$ are **logically equivalent** if one is true if and only if the other is true. We represent this symbolically by $P \iff Q$. For example

| $P$ | $\neg P$ | $\neg(\neg P)$ |
|---|---|---|
| 0 | 1 | 0 |
| 1 | 0 | 1 |

so for any statement,

$$\neg(\neg P) \iff P \tag{7.1}$$

which is sometimes called the Principle of Double Negation. We can use truth tables to verify many other logical equivalences. For example:

| $P$ | $Q$ | $P \wedge Q$ | $\neg(P \wedge Q)$ | $\neg P$ | $\neg Q$ | $(\neg P) \vee (\neg Q)$ |
|---|---|---|---|---|---|---|
| 0 | 0 | 0 | 1 | 1 | 1 | 1 |
| 0 | 1 | 0 | 1 | 1 | 0 | 1 |
| 1 | 0 | 0 | 1 | 0 | 1 | 1 |
| 1 | 1 | 1 | 0 | 0 | 0 | 0 |

from which we deduce that, for all statements $P, Q$,

$$\neg(P \wedge Q) \iff (\neg P) \vee (\neg Q). \tag{7.2}$$

As a consequence of (7.1) and (7.2),

$$\neg(P \vee Q) \iff \neg[(\neg\neg P) \vee (\neg\neg Q)] \quad \text{by (7.1)}$$
$$\iff \neg[\neg((\neg P) \wedge (\neg Q))] \quad \text{by (7.2)}$$
$$\iff (\neg P) \wedge (\neg Q) \quad \text{by (7.1)}$$

that is, for all statements $P, Q$,

$$\neg(P \vee Q) \iff (\neg P) \wedge (\neg Q). \tag{7.3}$$

**Exercise 7.1.** Verify this explicitly using a truth table.

The rules (7.2) and (7.3) are known as **de Morgan's laws**.

**Example 7.2.** Find the negation of the statement "Reality TV is the work of the devil, or $\sqrt{2}$ is rational and I'm a Dutchman."
*Solution:* Let's introduce symbols $P$, $Q$, $R$ for the statements

$P$: Reality TV is the work of the devil
$Q$: $\sqrt{2}$ is rational
$R$: I'm a Dutchman

Then the statement we're asked to negate is $P \vee (Q \wedge R)$. Now

$$\neg[P \vee (Q \wedge R)] \iff (\neg P) \wedge \neg(Q \wedge R) \qquad \text{by (7.3)}$$
$$\iff (\neg P) \wedge [(\neg Q) \vee (\neg R)] \qquad \text{by (7.2)}$$

Translating back to ordinary English, the negation is

"Reality TV is not the work of the devil, and $\sqrt{2}$ is irrational or I'm not a Dutchman." $\qquad \square$

## 7.2 Implications

It is convenient to define another connective between two statements $P, Q$ in addition to conjunction $(P \wedge Q)$ and disjunction $(P \vee Q)$, namely **implication**, denoted $P \Rightarrow Q$, pronounced "if $P$ then $Q$", or "$P$ implies $Q$". (Actually, the standard notation is $P \rightarrow Q$, and $P \Rightarrow Q$ is reserved for something subtly different. However, we are already making heavy use of the symbol $\rightarrow$ to denote convergence of sequences, so we will stick to $\Rightarrow$ for implication.) The truth table for this statement is

| $P$ | $Q$ | $P \Rightarrow Q$ |
|:---:|:---:|:---:|
| 0 | 0 | 1 |
| 0 | 1 | 1 |
| 1 | 0 | 0 |
| 1 | 1 | 1 |

As an example, consider the statement

"If everyone in my next real analysis class passes the final exam, I will eat my hat."

This is clearly false if everyone passes the exam and I refuse to eat my hat. On the other hand, it's clearly true if everyone passes and I *do* eat my hat. What if not everyone passes? Then the question of whether I eat my hat becomes moot: the statement is true if I don't, but it's also true if I do (I never said I'd eat my hat *only* if everyone passes – maybe I just like eating hats!). So $P \Rightarrow Q$ is false only when $P$ is true and $Q$ is false. In every other case, it is true.

Unlike $P \wedge Q$ and $P \vee Q$, the order of the statements $P, Q$ in $P \Rightarrow Q$ matters, as can be seen immediately from its truth table. For example, the statement

"If you are vegetarian then you don't eat pork"

is true, whereas

"If you don't eat pork then you are vegetarian"

may well be false. Given an implication $P \Rightarrow Q$, the implication $Q \Rightarrow P$ is called its **converse**. As we have just seen, an implication and its converse are *not* logically equivalent. We have seen many examples of this fact in this book. For example

"If a real sequence converges then it is bounded"

is true (see Proposition 3.3), but its converse

"If a real sequence is bounded then it converges"

is false (e.g. $a_n = (-1)^n$). Similarly

"If a real sequence is bounded then it has a convergent subsequence"

is true (this is the Bolzano–Weierstrass Theorem), but its converse

"If a real sequence has a convergent subsequence then it is bounded"

is false (e.g. $a_n = (1 + (-1)^n)n$). Another important, and often misused, example is

"If $\sum_{n=1}^{\infty} a_n$ converges then $a_n \to 0$"

which is true (this is the Divergence Test), but its converse

"If $a_n \to 0$ then $\sum_{n=0}^{\infty} a_n$ converges"

is false (e.g. $\sum_{n=1}^{\infty} 1/n$).

It is possible to rewrite $P \Rightarrow Q$ using only $\neg$ and $\vee$, and this is often useful when analyzing the precise meaning of an implication. In fact

$$[P \Rightarrow Q] \quad \Longleftrightarrow \quad [(\neg P) \vee Q]. \tag{7.4}$$

We can check this by computing the truth table of the right-hand side:

| $P$ | $Q$ | $\neg P$ | $(\neg P) \vee Q$ |
|---|---|---|---|
| 0 | 0 | 1 | 1 |
| 0 | 1 | 1 | 1 |
| 1 | 0 | 0 | 0 |
| 1 | 1 | 0 | 1 |

From this, we deduce that

$$[P \Rightarrow Q] \iff [(\neg P) \vee Q]$$
$$\iff [Q \vee (\neg P)]$$
$$\iff [(\neg(\neg Q)) \vee (\neg P)] \qquad \text{by (7.1)}$$
$$\iff [(\neg Q) \Rightarrow (\neg P)] \qquad \text{by (7.4)}$$

So, for any pair of statements $P, Q$,

$$[P \Rightarrow Q] \iff [(\neg Q) \Rightarrow (\neg P)].$$

**Exercise 7.3.** Check this by computing the truth table of $(\neg Q) \Rightarrow (\neg P)$.

Given an implication $P \Rightarrow Q$, the implication $(\neg Q) \Rightarrow (\neg P)$ is called its **contrapositive**. We have just shown that an implication and its contrapositive are logically equivalent. Looking back at the above examples, we see that Proposition 3.3 could equally well have been stated

"If a real sequence is unbounded then it does not converge",

and the Bolzano–Weierstrass Theorem is equivalent to

"If a real sequence has no convergent subsequence then it is unbounded",

and the Divergence Test is equivalent to

"If $a_n$ does not converge to 0 then $\sum_{n=1}^{\infty} a_n$ diverges",

which is, in fact, the form in which one actually uses it.

**Example 7.4.** Write down the converse and contrapositive of each of the following implications.

(i) *If $x^2 = 2$ then $x$ is irrational*

Converse: if $x$ is irrational then $x^2 = 2$
Contrapositive: if $x$ is rational then $x^2 \neq 2$

(ii) *If $a_n \to A$ and $b_n \to B$ then $a_n + b_n \to A + B$*

Converse: if $a_n + b_n \to A + B$ then $a_n \to A$ and $b_n \to B$
Contrapositive: if $a_n + b_n \not\to A + B$ then $a_n \not\to A$ or $b_n \not\to B$

(iii) *If $x > 2$ then $x^2 > 4$*

> Converse: If $x^2 > 4$ then $x > 2$
> Contrapostive: if $x^2 \leq 4$ then $x \leq 2$ □

If you are asked to prove an implication, you should always consider whether proving its contrapositive will be easier. Since the contrapositive is logically equivalent to the original implication, this will suffice.

**Example 7.5.** Claim: let $n$ be an integer. If $n^2$ is divisible by 3 then $n$ is divisible by 3.

**Proof.** We will prove the contrapositive: if $n$ is *not* divisible by 3 then $n^2$ is *not* divisible by 3. So, assume that $n$ is not divisible by 3. Then its remainder on division by 3 is either 1 or 2, that is, either $n = 3k + 1$ or $n = 3k + 2$, where $k$ is an integer. In the first case

$$n^2 = 9k^2 + 6k + 1 = 3k' + 1$$

where $k' = 3k^2 + 2k \in \mathbb{Z}$, and in the second case

$$n^2 = 9k^2 + 12k + 4 = 3k' + 1$$

where $k' = 3k^2 + 4k + 1 \in \mathbb{Z}$. In either case, we see that $n^2$ is not divisible by 3 (in fact, its remainder is 1 in both cases), as was to be proved. □

The last thing we should note about implications is how they behave under negation:

$$\neg[P \Rightarrow Q] \iff \neg[\neg P \vee Q] \iff [P \wedge \neg Q].$$

In words, the negation of "if $P$ then $Q$" is the statement "$P$ and not $Q$". Note in particular that the negation of an implication is *not* an implication.

## 7.3 Quantifiers

We have been a little bit sloppy up to now in dealing with statements which contain variables. Consider, for example, the statement

$$x > 2.$$

Is this really a statement? Until one specifies a particular value for $x$, one cannot say whether it is true. So it's true for $x = 3$ and false for $x = -2$, for example. What about if $x = 1 + 3i$? The "statement" $1 + 3i > 2$ is pretty meaningless, and it isn't very helpful to declare it either true or false (e.g. if you decide it's false, then you'd expect $1 + 3i \leq 2$ to be true, and this "statement" is equally meaningless). If we're being careful, we should make explicit the condition that $x$ is a real number, and if we denote the statement $x > 2$ with a symbol, we should choose one like $P(x)$, rather than $P$, to remind ourselves that whether $P(x)$ is true depends on the particular $x$ under consideration. So $P(3)$ is true, in this case, while $P(2)$ is false.

There are two very useful ways to turn a statement with a variable into an unambiguous true-or-false statement, using **quantifiers**. Let $P(x)$ be any statement which makes sense for all values of $x$ in some specified set $A$ (e.g. $A = \mathbb{R}$ and $P(x) : x > 2$, as above). Then

$\forall x \, P(x)$ means "for all $x \in A$, $P(x)$ is true"

and

$\exists x \, P(x)$ means "there exists some $x \in A$ such that $P(x)$ is true".

The symbol $\forall$ is called the **universal quantifier** and the symbol $\exists$ is called the **existential quantifier**. For the example above, $\forall x \, P(x)$ is false, because, for example, $P(-1)$ is false (since $-1$ is not greater than 2), but $\exists x \, P(x)$ is true because, for example, $P(3)$ is true. Note that, once we've coupled it with a quantifier, the statement no longer contains a variable and is a simple true or false assertion. Note also that it's entirely irrelevant what we *call* the variable. So $\forall y \, P(y)$ is precisely the same statement as $\forall x \, P(x)$.

In practice, rather than specifying the set $A$ in advance, we often specify it explicitly with the quantifier. In the above example, we would write

$$\forall x \in \mathbb{R}, \, x > 2$$

which is false, and

$$\exists x \in \mathbb{R}, \, x > 2 \tag{7.5}$$

which is true. You should feel free to add punctuation and/or ordinary English to expressions like this, if you think it clarifies the meaning. For example, I think

$$\exists x \in \mathbb{R} \text{ such that } x > 2$$

is clearer than (7.5). Indeed, you should feel free to avoid using the symbols $\forall$ and $\exists$ entirely if you think they obscure the mathematical meaning of the sentences you are writing. As we will see, they can be extremely useful, but you shouldn't chuck them into your mathematical prose willy-nilly. They won't make your proofs any more rigorous, elegant, or impressive.

**Example 7.6.** Translate the following sentences into symbols:

(i) The square of any even integer is even.

In symbols: $\forall n \in \mathbb{Z}, [(n/2) \in \mathbb{Z} \Rightarrow (n^2/2) \in \mathbb{Z}]$

(ii) Given any real number $K$ there is a positive integer $N$ such that $N > K$.

In symbols: $\forall K \in \mathbb{R}, \exists N \in \mathbb{Z}^+, N > K$

(iii) There exists a rational number whose square is 2.

In symbols: $\exists x \in \mathbb{Q}, x^2 = 2$

(iv) There exists a real number which is bigger than all rational numbers.

In symbols: $\exists x \in \mathbb{R}, \forall y \in \mathbb{Q}, x > y$

(v) The sequence $(a_n)$ converges to $L$:

In symbols: $\forall \varepsilon \in (0, \infty), \exists N \in \mathbb{Z}^+, \forall n \in \mathbb{Z}^+ \cap [N, \infty),$
$|a_n - L| < \varepsilon$

(vi) The sequence $(a_n)$ converges.

In symbols: $\exists L \in \mathbb{R}, \forall \varepsilon \in (0, \infty), \exists N \in \mathbb{Z}^+, \forall n \in \mathbb{Z}^+ \cap [N, \infty),$
$|a_n - L| < \varepsilon$　　　　　　　　□

**Example 7.7.** Translate the following "sentences" into English:

(i) $\forall x \in \mathbb{Z}, x^2 > x$

Translation: the square of every integer $x$ exceeds $x$.

Better translation: every integer is less than its square.

(ii) $\exists x \in \mathbb{R}, \forall y \in \mathbb{R}, x^2 > y$

Translation: there exists a real number whose square is greater than every real number.

(iii) $\forall y \in \mathbb{R}, \exists x \in \mathbb{R} \ x^2 > y$

Translation: for each real number $y$, there exists a real number $x$ whose square exceeds $y$.

Better translation: the function $f : \mathbb{R} \to \mathbb{R}$, $f(x) = x^2$, is unbounded above. □

In Example 7.7, we had a statement depending on two variables, $P(x, y)$, to which we applied a pair of quantifiers, $\exists x$ and $\forall y$. An important point to notice is that *it matters which order we apply the quantifiers in!* That is

$$\exists x \ \forall y \ P(x, y) \qquad \text{is not the same as} \qquad \forall y \ \exists x \ P(x, y).$$

In Example 7.7, we had $P(x, y) : x^2 > y$, from which we constructed

$$S_1 : \qquad \exists x \in \mathbb{R}, \forall y \in \mathbb{R}, P(x, y) \qquad\qquad \text{FALSE!}$$

and

$$S_2 : \qquad \forall y \in \mathbb{R}, \exists x \in \mathbb{R}, P(x, y) \qquad\qquad \text{TRUE!}$$

Since $S_2$ is true and $S_1$ is false, they're clearly not equivalent.

**Example 7.8.** Here's a non-mathematical example illustrating the same point. Let $X$ be the set of all people (in your family, say), $C$ be the set of all colours, and $P(x, c)$ be the statement "$x$'s favourite colour is $c$". Then

$$\forall x \in X, \ \exists c \in C, \ P(x, c) \qquad \Longleftrightarrow \qquad \text{Everyone has a favourite colour,}$$

whereas

$$\exists c \in C, \ \forall x \in X, \ P(x, c) \qquad \Longleftrightarrow \qquad \text{There is a colour which is everyone's favourite,}$$

which is clearly a very different statement. □

The situation in which quantifiers really come into their own is when you want to understand the *negation* of a statement involving several variables. Let's ask ourselves, what does it mean to say that $\forall x \ P(x)$ is *false*? It means that it is *not* true that for all $x$ in $A$, $P(x)$ is true. That is, there is at least one element $x$ in $A$ for which $P(x)$ is *not* true. In other words, there exists $x \in A$ such that $\neg P(x)$ is true. That is

$$\neg [\forall x \ P(x)] \qquad \Longleftrightarrow \qquad \exists x \ \neg P(x). \qquad (7.6)$$

**Example 7.9.** Consider the assertion that "all politicians are corrupt". I think this is false. In saying that, I am certainly *not* asserting that "all politicians are not corrupt" (which is equivalent to "no politician is corrupt"). Rather, I am asserting that "there exists at least one politician who is not corrupt". In symbols, let $X$ denote the set of politicians and $P(x)$ be the statement "$x$ is corrupt". Then the negation of $\forall x \in X\, P(x)$ is *not* $\forall x \in X\, \neg P(x)$, but rather $\exists x \in X\, \neg P(x)$. □

Note that to prove that $[\forall x \in A\, P(x)]$ is *false*, we need to show that $[\exists x \in A\, \neg P(x)]$ is true, that is, we need to show that there exists (at least) *one* element $x$ in $A$ such that $P(x)$ is *false*. Such an element $x$ is called a **counterexample** to $[\forall x \in A\, P(x)]$.

**Example 7.10.** Consider the statement $\forall x \in \mathbb{R}\, x^2 < 10$. This is clearly false. To prove that it is false, all we need to do is exhibit a *single* real number whose square is not less than 10. For example, $x = 10$ will do (since $10^2 \geq 10$). To prove the statement is false we do **not** have to find *all* real numbers $x$ such that $x^2 \geq 10$. Giving a single counterexample is quicker, clearer, and more elegant. □

The rule for negating $\exists x\, P(x)$ actually follows from the logical equivalence (7.6) and rule (7.1), but let's derive it by thinking. What does it mean to say that $\exists x\, P(x)$ is *false*? It means that there does *not* exist $x$ in $A$ such that $P(x)$ is true. Hence, for each and every $x \in A$, $P(x)$ must be *false*. That is

$$\neg[\exists x\, P(x)] \qquad \Longleftrightarrow \qquad \forall x\, \neg P(x).$$

**Example 7.11.** Consider the assertion "someone in my family speaks fluent Mandarin". Let's assume this is false. What does this mean? It certainly does *not* merely mean "someone in my family does not speak fluent Mandarin". Rather it means "no-one in my family speaks fluent Mandarin", or, equivalently, "everyone in my family does not speak fluent Mandarin". In symbols, let $X$ denote the set of people in my family and $P(x)$ be the statement "$x$ speaks fluent Mandarin". Then the negation of $\exists x \in X,\, P(x)$ is *not* $\exists x \in X,\, \neg P(x)$, but rather $\forall x \in X,\, \neg P(x)$. □

**Example 7.12.** Negate the following statements. Determine which is true, the statement or its negation.

(i)

$$S: \qquad \forall x \in \mathbb{Z},\, x^2 > x$$
$$\neg S: \qquad \exists x \in \mathbb{Z},\, x^2 \leq x$$

$\neg S$ is true (for example $0^2 \leq 0$).

(ii)

$$S: \qquad \exists x \in \mathbb{R}, \ x^2 + 2x + 2 = 0$$
$$\neg S: \qquad \forall x \in \mathbb{R}, \ x^2 + 2x + 2 \neq 0$$

$\neg S$ is true, since $x^2 + 2x + 2 = (x+1)^2 + 1 \geq 1$.

(iii)

$$S: \qquad \forall x \in (0, \infty), \ \exists y \in \mathbb{Q}, \ 0 < y < 1/x$$
$$\neg S: \qquad \exists x \in (0, \infty), \ \neg[\exists y \in \mathbb{Q}, \ 0 < y < 1/x]$$
$$\Longleftrightarrow \ \exists x \in (0, \infty), \ \forall y \in \mathbb{Q}, \ \neg[0 < y < 1/x]$$
$$\Longleftrightarrow \ \exists x \in (0, \infty), \ \forall y \in \mathbb{Q}, \ \neg[0 < y \wedge y < 1/x]$$
$$\Longleftrightarrow \ \exists x \in (0, \infty), \ \forall y \in \mathbb{Q}, \ [y \leq 0 \vee y \geq 1/x]$$

$S$ is true by Theorem 1.25: given any $x > 0$, $1/x > 0$, so there exists a rational number $y$ between 0 and $1/x$.                                 □

You've already seen these negation rules in many different contexts. For example, consider the Archimedean Property of $\mathbb{R}$. Our first statement of this property was simply that "$\mathbb{Z}^+$ is unbounded above" (see Example 1.17). We then reformulated this as

"Given any real number $K$, there is some positive integer $n$ such that $n > K$."

We can now see symbolically that these two statements are logically equivalent:

$$\mathbb{Z}^+ \text{ is unbounded above} \Longleftrightarrow \neg[\mathbb{Z}^+ \text{ is bounded above}]$$
$$\Longleftrightarrow \neg[\exists K \in \mathbb{R}, \ \forall n \in \mathbb{Z}^+, \ n \leq K]$$
$$\Longleftrightarrow \forall K \in \mathbb{R}, \neg[\forall n \in \mathbb{Z}^+, \ n \leq K]$$
$$\Longleftrightarrow \forall K \in \mathbb{R}, \ \exists n \in \mathbb{Z}^+, \ \neg[n \leq K]$$
$$\Longleftrightarrow \forall K \in \mathbb{R}, \ \exists n \in \mathbb{Z}^+, \ n > K.$$

In this case, we could get to the end result just by thinking carefully about what the phrase "$\mathbb{Z}^+$ is unbounded above" really means (as we did back in section 1.3). Sometimes, when you're dealing with the negation of a subtle statement with lots of quantifiers, it really does clarify the situation to write things out symbolically and use the negation rules. A good example is the statement "$(a_n)$ does not converge to $L$". To illustrate, consider the following problem:

**Example 7.13.** Let $(a_n)$ be a real sequence with the property that every subsequence of $(a_n)$ has a subsequence which converges to 0. Show that $(a_n)$ converges to 0.

**Proof.** We will prove this by contradiction. So, let $(a_n)$ be a sequence with the specified property, but assume that $(a_n)$ does *not* converge to 0. Our first task is to understand precisely what this means:

$$
\begin{aligned}
a_n \nrightarrow 0 &\iff \neg[a_n \to 0] \\
&\iff \neg[\forall \varepsilon \in (0, \infty),\ \exists N \in \mathbb{Z}^+,\ \forall n \in \mathbb{Z}^+ \cap [N, \infty,)\ |a_n - 0| < \varepsilon] \\
&\iff \exists \varepsilon \in (0, \infty),\ \neg[\exists N \in \mathbb{Z}^+,\ \forall n \in \mathbb{Z}^+ \cap [N, \infty),\ |a_n| < \varepsilon] \\
&\iff \exists \varepsilon \in (0, \infty),\ \forall N \in \mathbb{Z}^+,\ \neg[\forall n \in \mathbb{Z}^+ \cap [N, \infty),\ |a_n| < \varepsilon] \\
&\iff \exists \varepsilon \in (0, \infty),\ \forall N \in \mathbb{Z}^+,\ \exists n \in \mathbb{Z}^+ \cap [N, \infty),\ \neg[|a_n| < \varepsilon] \\
&\iff \exists \varepsilon \in (0, \infty),\ \forall N \in \mathbb{Z}^+,\ \exists n \in \mathbb{Z}^+ \cap [N, \infty),\ |a_n| \geq \varepsilon.
\end{aligned}
$$

Translating back into English, $a_n \nrightarrow 0$ means

"There exists $\varepsilon > 0$ such that, for all $N \in \mathbb{Z}^+$ there is some integer $n \geq N$ such that $|a_n| \geq \varepsilon$."

Let $\varepsilon_*$ be this magic positive real number. Then we know that, for each and every $N \in \mathbb{Z}^+$, there is some $n \geq N$ such that $|a_n| \geq \varepsilon_*$. So this is true for $N = 1$, for example: there exists an integer, $n_1$ say, such that $n_1 \geq 1$ and $|a_{n_1}| \geq \varepsilon_*$. It's also true for $N = n_1 + 1$: there exists $n_2 \geq n_1 + 1$ such that $|a_{n_2}| \geq \varepsilon_*$. It's also true for $N = n_2 + 1$, and so on. In this way we generate a *subsequence* $b_k = a_{n_k}$ of $(a_n)$ such that $|b_k| = |a_{n_k}| \geq \varepsilon_*$ for all $k \in \mathbb{Z}^+$. By assumption, the subsequence $(b_k)$ itself has a subsequence which *does* converge to 0, $c_m = b_{k_m}$ say. Since $c_m \to 0$, given any postive number, for example our magic number $\varepsilon_*$, there is some positive integer $M$ such that, for all $m \geq M$, $|c_m - 0| < \varepsilon_*$. In particular, $|c_M| < \varepsilon_*$. But $c_M = b_{k_M}$, and the sequence $(b_k)$ was chosen specifically so that $|b_k| \geq \varepsilon_*$ for all $k \in \mathbb{Z}^+$, including $k_M$. Hence, we have reached a contradiction, so our intial assumption that $a_n \nrightarrow 0$ must be false. $\qquad\square$

**Remark.** This result is quite subtle. We can represent it symbolically as follows. Let $X$ be the set of all real sequences, and for each $a \in X$, let $X_a$ be the set of all subsequences of $a$. Clearly $X_a \subset X$. For each $b \in X$, let $P(b)$ be the statement "$b$ converges to 0". Then we have just proved

$$\forall a \in X,\ [\forall b \in X_a,\ \exists c \in X_b,\ P(c)] \Rightarrow P(a).$$

Here are a couple of questions for you to think about.

(i) Consider the statement

> "If every subsequence of $(a_n)$ has a subsequence which has a subsequence which converges to 0 then $(a_n)$ converges to 0."

We can translate this into symbols as

$$\forall a \in X, \ [\forall b \in X_a, \ \exists c \in X_b, \ \exists d \in X_c, \ P(d)] \Rightarrow P(a).$$

Is this true?

(ii) In Example 7.13, we assumed that all subsequences have a subsequence converging to 0. Presumably there's nothing special about 0. We could equally well have assumed that all subsequences have a subsequence converging to $L$ for any other fixed $L \in \mathbb{R}$. (If this isn't clear, try adapting the above proof to deal with the case of general $L$.) Now consider the following statement:

> "If every subsequence of $(a_n)$ has a convergent subsequence then $(a_n)$ is convergent."

Symbolically, for each $a \in X$ we can let $Q(a)$ be the statement "$a$ converges". Then the above statement is

$$\forall a \in X, \ [\forall b \in X_a, \ \exists c \in X_b, \ Q(c)] \Rightarrow Q(a).$$

Is this true?

## 7.4 Summary

- We can represent statements by symbols $P$, $Q$ etc. We say that $P$ and $Q$ are **logically equivalent** if $P$ is true if and only if $Q$ is true, denoted $P \iff Q$.
- We obtain new statements from existing ones using the following operations:
  - $\neg P$ denotes the **negation** of $P$, "not $P$", which is true if and only if $P$ is false
  - $P \wedge Q$ denotes the **conjunction** of $P$ and $Q$, "$P$ and $Q$", which is true if and only if both $P$ and $Q$ are true
  - $P \vee Q$ denotes the **disjunction** of $P$ and $Q$, "$P$ or $Q$", which is true if $P$ is true or $Q$ is true, or both.
- These operations obey algebraic rules:

$$\neg(\neg P) \iff P,$$
$$\neg(P \wedge Q) \iff (\neg P) \vee (\neg Q),$$
$$\neg(P \vee Q) \iff (\neg P) \wedge (\neg Q).$$

- $P \Rightarrow Q$ denotes the **implication** "if $P$ then $Q$". By definition, this is logically equivalent to $(\neg P) \vee Q$.
  - The **converse** of $P \Rightarrow Q$ is $Q \Rightarrow P$.
  - The **contrapositive** of $P \Rightarrow Q$ is $(\neg Q) \Rightarrow (\neg P)$.

  An implication is logically equivalent to its contrapositive but **not** to its converse.
- The symbols $\forall$ and $\exists$ are the **universal** and **existential quantifiers**.
  - $\forall x \in X$, $P(x)$ means "for all $x \in X$, $P(x)$ is true".
  - $\exists x \in X$, $P(x)$ means "there exists some $x \in X$ such that $P(x)$ is true".
- These symbols behave under negation as follows

$$\neg[\forall x \in X, \ P(x)] \iff \exists x \in X, \ \neg P(x),$$
$$\neg[\exists x \in X, \ P(x)] \iff \forall x \in X, \ \neg P(x).$$

## 7.5    Tutorial problems

1. Write down the truth table for the statement $P \Rightarrow (Q \wedge (R \Rightarrow \neg P))$.
2. Translate the following English sentences into symbols.

    (a) Every rational number is a quotient of two irrational numbers.
    (b) The set $A = \{n + 1/n : n \in \mathbb{Z}^+\}$ is bounded below but unbounded above.
    (c) The function $f : \mathbb{R} \to \mathbb{R}$, $f(x) = x^3 - x$ is injective.

    Which, if any, of these statements are true?
3. Translate the following symbolic sentences into English.

    (a) $\forall(x, y) \in \mathbb{R}^2$, $x < y \Rightarrow \exists z \in \mathbb{Q}$, $x < z < y$.
    (b) $\forall y \in \mathbb{R}$, $\exists x \in \mathbb{R}$, $x^3 - x = y$.
4. Negate the following statements and determine which is true, the statement or its negation.

    (a) $P : \forall x \in \mathbb{R}$, $x^2 > x \Rightarrow x^4 > x^2$.
    (b) $Q : \forall x \in \mathbb{R}$, $\exists y \in \mathbb{R}$, $|y| = x + y$.
    (c) $R : \exists x \in \mathbb{R}$, $\forall y \in \mathbb{R}$, $x + y < |y|$.

## 7.6    Homework problems

1. Write down the truth table for the statement
   $(P \Rightarrow (Q \vee R)) \Rightarrow (Q \wedge (R \Rightarrow \neg P))$.
2. Translate the following English sentences into symbols.

    (a) Every real number has a cube root.
    (b) The sequence $2^n$ is unbounded above.
    (c) The function $f : \mathbb{R} \to \mathbb{C}$, $f(x) = x^2 + ix$, is not injective.

    Which, if any, of these statements are true?
3. Translate the following symbolic sentences into English. Make your answer as concise as possible.

    (a) $\forall n \in \mathbb{Z}^+$, $2^{n+1} \leq 2^n$.
    (b) $\forall x \in \mathbb{R}$, $\exists y \in \mathbb{Q}$, $y > x$.
    (c) $\exists a \in \mathbb{R}$, $\forall x \in \mathbb{R}$, $x + |x| \neq a$.

    Which, if any, of these statements are true?
4. Negate the following statements and determine which is true, the statement or its negation.

    (a) $P : \exists K \in \mathbb{R}$, $\forall n \in \mathbb{Z}^+$, $1/n \geq K$.

(b) $Q : \forall x \in \mathbb{R},\ \exists \varepsilon \in (0, \infty),\ \forall y \in (x - \varepsilon, x + \varepsilon),\ y \neq 1.$

(c) $R : \forall z \in \mathbb{C},\ |z| < 1 \Rightarrow |z| < |z - i|.$

# Chapter 8

# Limits of functions

## 8.1 The main definition

We now have a very precise (and powerful) definition of *limit* for real sequences. In this chapter we will show how to use this to give a precise definition of the limit of a real function of a real variable. We will also see that basic properties of limits of functions, which are usually asserted without proof in introductory calculus courses, follow very easily from the corresponding limit theorems for sequences, which we proved in section 3.1.

Informally, $\lim_{x \to a} f(x) = L$ means that $f(x)$ is "arbitrarily close" to $L$ for all $x$ "sufficiently close to, but different from" $a$. So limits concern the behaviour of $x$ arbitrarily close to $a$, but *not at* $a$. There is no reason why $a$ has to be in the domain of $f$, that is, $f(a)$ may, or may not, be defined. If it is, its value should be irrelevant to $\lim_{x \to a} f(x)$. For example,

$$f(x) = \frac{\sin x}{x}$$

is clearly undefined at 0 (its domain is $(-\infty, 0) \cup (0, \infty)$), but $\lim_{x \to 0} f(x)$ is well-defined (it's 1).

In order to probe the behaviour of $f(x)$ for values of $x$ close to, but different from, $a$, we can consider the sequence $f(x_n)$ where $(x_n)$ is any sequence *converging* to $a$ with $x_n \neq a$. This leads us to the main definition of this chapter:

**Definition 8.1.** Let $D \subseteq \mathbb{R}$ and $f : D \to \mathbb{R}$. Then $f$ has **limit** $L$ at $a$ if, for all sequences $(x_n)$ in $D \backslash \{a\}$ such that $x_n \to a$, $f(x_n) \to L$. In this case, we usually denote the number $L$ by $\lim_{x \to a} f(x)$.

It's important to note that the definition says that for *all* sequences $(x_n)$ in $D \backslash \{a\}$ converging to $a$, $f(x_n) \to L$. It's not enough to show that

$f(x_n) \to L$ for just one specific sequence. Note that the sequence $(x_n)$ gets arbitrarily close to $a$ (since $x_n \to a$) but never *equals* $a$ (since it takes values in $D\backslash\{a\}$). So the value of $f$ at $a$, if this exists, is irrelevant.

In order for Definition 8.1 to make sense, there must *exist* at least one sequence in $D\backslash\{a\}$ such that $x_n \to a$. For example, the function

$$f(x) = \sqrt{1 - x^2}$$

has maximal domain $D = [-1, 1]$, so it doesn't make sense to try to define $\lim_{x\to 2} f(x)$, its limit at 2 say, because there don't exist any sequences in $D\backslash\{2\} = [-1, 1]$ which converge to 2. To be precise, we should include this requirement in our definition of limit. This leads us to the following definition.

**Definition 8.2.** Let $D \subset \mathbb{R}$ and $a \in \mathbb{R}$. We say that $a$ is a **cluster point** of $D$ if there exists a sequence $(x_n)$ in $D\backslash\{a\}$ such that $x_n \to a$.

**Example 8.3.** Claim: every $a \in \mathbb{R}$ is a cluster point of $\mathbb{Q}$.

**Proof.** Choose any $a \in \mathbb{R}$. By Theorem 1.25, for each $n \in \mathbb{Z}^+$ there exists a rational number $x_n$ between $a$ and $a + 1/n$. Now $x_n \in \mathbb{Q}\backslash\{a\}$ and $x_n \to a$ by the Squeeze Rule. Hence $a$ is a cluster point of $\mathbb{Q}$.  □

**Example 8.4.** Claim: the set of cluster points of $(0, \infty)$ is $[0, \infty)$.

**Proof.** Let $a \in [0, \infty)$. Then, for each $n \in \mathbb{Z}^+$, $x_n = a + \frac{1}{n}$ is different from $a$ and lies in $(0, \infty)$. Clearly $x_n \to a$. Hence every $a \in [0, \infty)$ is a cluster point of $(0, \infty)$, that is, the set of cluster points of $(0, \infty)$ contains $[0, \infty)$. Conversely, assume $a \notin [0, \infty)$. Then $a < 0$. Let $(x_n)$ be any sequence in $(0, \infty)$. If $(x_n)$ converges at all, its limit must be non-negative, since $x_n > 0$ for all $n$ (Proposition 3.4). Hence $x_n \not\to a$. Since no sequence in $(0, \infty)$ converges to $a$, $a$ is *not* a cluster point of $(0, \infty)$.  □

Many authors call cluster points **limit points** or **accumulation points**. It makes sense to ask whether $f : D \to \mathbb{R}$ has a limit at $a$ precisely when $a$ is a cluster point of $D$. With this is mind, we should really tighten up Definition 8.1.

**Improved Definition 8.1.** Let $D \subseteq \mathbb{R}$, $f : D \to \mathbb{R}$ and $a$ be a cluster point of $D$. Then $f$ has **limit** $L$ **at** $a$ if, for all sequences $(x_n)$ in $D\backslash\{a\}$

such that $x_n \to a$, $f(x_n) \to L$. In this case, we usually denote the number $L$ by $\lim_{x \to a} f(x)$.

An important fact about limits is that, if they exist, they're unique:

**Proposition 8.5.** *If* $\lim_{x \to a} f(x)$ *exists, then it is unique.*

**Proof.** Let $f : D \to \mathbb{R}$ and assume, to the contrary, that $K$ and $L$ are non-equal real numbers satisfying the definition of $\lim_{x \to a} f(x)$. Let $(x_n)$ be a sequence in $D \backslash \{a\}$ such that $x_n \to a$. Then $f(x_n) \to K$ and $f(x_n) \to L$ and $f(x_n) \to L \neq K$. But this contradicts Proposition 3.1. □

So it makes sense to speak of *the* limit of a function at $a$.

**Example 8.6.** Let $f : \mathbb{R} \backslash \{1\} \to \mathbb{R}$ such that

$$f(x) = \frac{x^2 - 1}{x - 1}.$$

Let $x_n$ be any sequence in $\mathbb{R} \backslash \{1\}$ converging to 1. Then

$$f(x_n) = \frac{(x_n - 1)(x_n + 1)}{x_n - 1} = x_n + 1 \to 2.$$

So $\lim_{x \to 1} f(x) = 2$, as you would expect. □

**Example 8.7.** Let $f : \mathbb{R} \backslash \{0\} \to \mathbb{R}$ such that

$$f(x) = \frac{x}{|x|}.$$

Then $\lim_{x \to 0} f(x)$ does not exist. To prove this, note that $x_n = (-1)^n / n$ is a sequence converging to 0, but

$$f(x_n) = \frac{(-1)^n / n}{1/n} = (-1)^n$$

which does not converge at all. □

**Example 8.8.** Let $f : \mathbb{R} \backslash \{0\} \to \mathbb{R}$ such that

$$f(x) = \sin \frac{1}{x}.$$

We can prove that $\lim_{x \to 0} f(x)$ does not exist using a similar argument. Assume that $\lim_{x \to 0} f(x)$ exists, and call it $L$. Let

$$x_n = \frac{1}{\pi n}.$$

Then $(x_n)$ is a sequence in $D\backslash\{0\} = D$ converging to 0, so $f(x_n) \to L$. But

$$f(x_n) = \sin \pi n = 0 \to 0,$$

so $L = 0$. Let

$$y_n = \frac{1}{\pi(2n + \frac{1}{2})}.$$

Again, this is a sequence in $D\backslash\{0\}$ converging to 0, so $f(y_n) \to L = 0$. But

$$f(y_n) = \sin\left(\frac{\pi}{2} + 2n\pi\right) = 1 \to 1,$$

a contradiction. $\qquad\qquad\qquad\qquad\qquad\qquad\qquad\qquad\qquad\qquad\square$

Note that $\lim_{x \to a} f(x)$ is a single number, not a function, and the fact that we called the original variable $x$ is of no significance. That is

$$\lim_{x \to a} f(x) = \lim_{y \to a} f(y) = \lim_{\widetilde{\Gamma}_0' \to a} f(\widetilde{\Gamma}_0').$$

If we have a function of two variables, $f(x, y)$, we can imagine holding one of them fixed, $y$ say, and taking a limit with respect to the other, $x$. The result is a function of $y$

$$F(y) = \lim_{x \to a} f(x, y).$$

We can then take a limit of $F$ with respect to $y$,

$$\lim_{y \to b} F(y) = \lim_{y \to b} \left(\lim_{x \to a} f(x, y)\right).$$

On the other hand, we could take the limits in the opposite order, that is, define a function

$$G(x) = \lim_{y \to b} f(x, y)$$

and then compute *its* limit as $x$ tends to $a$,

$$\lim_{x \to a} G(x) = \lim_{x \to a} \left(\lim_{y \to b} f(x, y)\right).$$

It is important to realize that these two limits, if they both exist, need *not* be equal.

**Example 8.9.** Let $f : \mathbb{R}^2 \backslash \{(0,0)\} \to \mathbb{R}$ such that $f(x, y) = \frac{x^2}{x^2 + y^2}$. Compare the two double limits

$$\lim_{y \to 0} \left(\lim_{x \to 0} f(x, y)\right), \qquad \lim_{x \to 0} \left(\lim_{y \to 0} f(x, y)\right).$$

*Solution:* Let $y$ be any fixed, non-zero real number, and define the function $f_1 : \mathbb{R} \to \mathbb{R}$, $f_1(x) = x^2/(x^2 + y^2)$. Given any sequence $(x_n)$ in $\mathbb{R}\backslash\{0\}$ such that $x_n \to 0$,

$$f_1(x_n) = \frac{x_n^2}{x_n^2 + y^2} \to \frac{0}{0 + y^2} = 0.$$

Hence, for all $y \neq 0$,

$$\lim_{x \to 0} f(x, y) = \lim_{x \to 0} f_1(x) = 0 =: F(y).$$

We want to compute the limit of this (constant) function as $y$ tends to 0. Given any sequence $(y_n)$ in $\mathbb{R}\backslash\{0\}$ such that $y_n \to 0$, $F(y_n) = 0 \to 0$. Hence $\lim_{y \to 0} F(y) = 0$. That is,

$$\lim_{y \to 0} \left( \lim_{x \to 0} f(x, y) \right) = 0.$$

Consider now the second double limit. Let $x$ be any fixed, non-zero real number, and define the function $f_2 : \mathbb{R} \to \mathbb{R}$, $f_2(y) = x^2/(x^2 + y^2)$. Given any sequence $(y_n)$ in $\mathbb{R}\backslash\{0\}$ such that $y_n \to 0$,

$$f_2(y_n) = \frac{x^2}{x^2 + y_n^2} \to \frac{x^2}{x^2 + 0} = 1.$$

Hence, for all $x \neq 0$,

$$\lim_{y \to 0} f(x, y) = \lim_{y \to 0} f_2(y) = 1 =: G(x).$$

We want to compute the limit of this (constant) function as $x$ tends to 0. Given any sequence $(x_n)$ in $\mathbb{R}\backslash\{0\}$ such that $x_n \to 0$, $G(x_n) = 1 \to 1$. Hence $\lim_{x \to 1} G(x) = 1$. That is,

$$\lim_{x \to 0} \left( \lim_{y \to 0} f(x, y) \right) = 1.$$

Clearly these two limits differ. □

Comparing Definition 8.1 with Definition 6.1, it's clear that there is a connexion between limits and continuity. The next proposition makes the link precise:

**Proposition 8.10.** *Let $f : D \to \mathbb{R}$, $a \in D$ and $a$ be a cluster point of $D$. Then $f$ is continuous at $a$ if and only if $\lim_{x \to a} f(x) = f(a)$.*

**Proof.** ($\Rightarrow$): Assume $f$ is continuous at $a$. Let $(x_n)$ be any sequence in $D\backslash\{a\}$ such that $x_n \to a$. Then $(x_n)$ is a sequence in $D$ such that $x_n \to a$. Since $f$ is continuous at $a$, $f(x_n) \to f(a)$. Hence $\lim_{x \to a} f(x) = f(a)$.

($\Leftarrow$): Assume that $\lim_{x \to a} f(x) = f(a)$. Let $(x_n)$ be any sequence in $D$ such that $x_n \to a$, and consider the set $A = \{n \in \mathbb{Z}^+ : x_n = a\}$. Either $A$ is bounded above or it is unbounded above. If $A$ is bounded above, then there exists $N \in \mathbb{Z}^+$ such that, for all $n \geq N$, $x_n \neq a$. Then the tail subsequence $y_n = x_{n+N}$ converges to $a$ and never equals $a$, so $f(y_n) \to f(a)$ (since, by assumption $\lim_{x \to a} f(x) = f(a)$). Hence, in this case, $f(x_n) \to f(a)$, by the Tail Lemma (Lemma 4.4). Consider now the case where $A$ is unbounded above. Then its complement $\mathbb{Z}^+\backslash A = \{n \in \mathbb{Z}^+ : x_n \neq a\}$ is bounded above or is unbounded above. If $\mathbb{Z}^+\backslash A$ is bounded above, then there exists $N \in \mathbb{Z}^+$ such that, for all $n \geq N$, $x_n = a$. But then, $f(x_{n+N}) = f(a) \to f(a)$, so $f(x_n) \to f(a)$ by Lemma 4.4.

Finally, we must deal with the case where both $A$ and $\mathbb{Z}^+\backslash A$ are unbounded above. In this case, we define the subsequences $(y_k) = (x_{m_k})$ where $m_k$ increases through all values in $A$, and $(z_k) = (x_{n_k})$ where $n_k$ increases through all values in $\mathbb{Z}^+\backslash A$. By construction, $(y_k), (z_k)$ is a covering pair for $(x_n)$ so, by Lemma 4.8, it suffices to show that $f(y_k) \to f(a)$ and $f(z_k) \to f(a)$. Now, by definition $y_k = a$, constant, so $f(y_k) = f(a) \to f(a)$. Furthermore, by definition, $z_k \neq a$ for all $k$ and $z_k \to a$ (as it's a subsequence of $(x_n)$). Hence $f(z_k) \to f(a)$ (since $\lim_{x \to a} f(x) = f(a)$). $\qquad\qquad\square$

**Example 8.11.** Let $f : \mathbb{R} \to \mathbb{R}$ such that $f(x) = x^{12} + 12x + 12$. Then $f$ is continuous at $-1$, so $\lim_{x \to -1} f(x) = f(-1) = 1$. $\qquad\qquad\square$

Having defined limits purely in terms of convergence of sequences, it is very easy to establish many of their basic properties, since these follow directly from theorems about sequences that we've already proved. We've already seen an example of this: limits of functions are unique (Proposition 8.5) because limits of convergent sequences are unique (Proposition 3.1). Another very useful example is the Algebra of Limits for functions:

**Theorem 8.12 (The Algebra of Limits for functions).**
Let $f : D \to \mathbb{R}$, $g : D \to \mathbb{R}$, $\lim_{x \to a} f(x) = L$, and $\lim_{x \to a} g(x) = K$. Then

(i) $\lim_{x \to a} (f(x) + g(x)) = L + K$,

(ii) $\lim_{x \to a} f(x)g(x) = LK$,

*and if, in addition, $f(x) \neq 0$ for all $x \in D\backslash\{a\}$ and $L \neq 0$, then*

*(iii)* $\displaystyle\lim_{x \to a} \frac{1}{f(x)} = \frac{1}{L}$.

**Proof.** Let $(x_n)$ be any sequence in $D\backslash\{a\}$ such that $x_n \to a$. Then $f(x_n) \to L$ and $g(x_n) \to K$ by assumption. Hence

(i) $f(x_n) + g(x_n) \to L + K$ by Theorem 3.5(i).
(ii) $f(x_n)g(x_n) \to LK$ by Theorem 3.5(ii).
(iii) $f(x_n) \neq 0$ and $L \neq 0$ so $1/f(x_n) \to 1/L$ by Theorem 3.5(iii).

$\square$

Note that the theorem follows almost immediately from Theorem 3.5 (the Algebra of Limits for sequences), once we've defined limits of functions in the right way (Definition 8.1).

## 8.2 Limits at infinity

Another important type of limit which one encounters informally in introductory calculus is $\lim_{x \to \infty} f(x)$. Again, we can make this precise using sequences. First, we have to define what it means for a sequence to "diverge to infinity".

**Definition 8.13.** A real sequence $(a_n)$ **diverges to infinity** if, for each $K \in \mathbb{R}$, there exists $N \in \mathbb{Z}$ such that $a_n > K$ for all $n \geq N$. In this case, we write $a_n \to \infty$. Similarly, $(a_n)$ **diverges to minus infinity** if, for each $K \in \mathbb{R}$, there exists $N \in \mathbb{Z}$ such that $a_n < K$ for all $n \geq N$. In this case, we write $a_n \to -\infty$.

## Remarks

- Having learned about quantifiers, we can now give a symbolic version of these definitions:

$$a_n \to \infty \quad \Leftrightarrow \quad \forall K \in \mathbb{R}, \exists N \in \mathbb{Z}^+, \forall n \in \mathbb{Z}^+ \cap [N, \infty), a_n > K,$$
$$a_n \to -\infty \quad \Leftrightarrow \quad \forall K \in \mathbb{R}, \exists N \in \mathbb{Z}^+, \forall n \in \mathbb{Z}^+ \cap [N, \infty), a_n < K.$$

- We can visualize these definitions geometrically, in the same way as we did for the definition of convergence. Imagine graphing the sequence $(a_n)$ with $n$ along the $x$-axis and $a_n$ along the $y$-axis. Assume $a_n \to \infty$.

Then, given any $K \in \mathbb{R}$, draw the horizontal line $y \doteq K$. The definition says that there is a point in the sequence (the $N$th term) to the right of which all points on the graph lie above the line $y = K$.

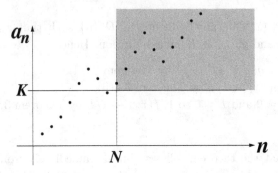

If we choose a different, larger $K$, the corresponding $N$ may need to be larger too. But no matter how big we make $K$, there is a point to the right of which all points on the graph lie above the line $y = K$.

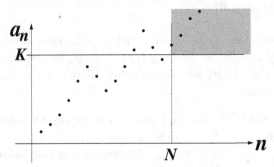

**Example 8.14.** Let $a_n = n$. Then $a_n \to \infty$ by the Archimedean Property of $\mathbb{R}$: given any $K \in \mathbb{R}$ there exists $N \in \mathbb{Z}^+$ such that $N > K$, and then for all $n \geq N$, $a_n = n \geq N > K$. $\qquad\square$

**Warning!** It's clear (hopefully) that if $a_n \to \infty$ then $(a_n)$ is unbounded above. (Given any $K \in \mathbb{R}$, there exists $N \in \mathbb{Z}^+$ such that $a_N > K$.) However the converse is false: not every sequence which is unbounded above diverges to infinity.

**Example 8.15.** Let $a_n = (-1)^n n$. This is certainly unbounded above. But it does not diverge to infinity because, for example, there is no $N$ such that $a_n > 0$ for all $n \geq N$ (because there are always odd numbers $n$ bigger than

$N$, and $a_n < 0$ whenever $n$ is odd). By similar reasoning, this sequence does *not* diverge to minus infinity, although it is unbounded below. □

The relationship between divergence to infinity and unboundedness above is slightly subtle.

**Proposition 8.16.** $a_n \to \infty$ *if and only if every subsequence of* $(a_n)$ *is unbounded above.*

**Proof.**
($\Rightarrow$) If $a_n \to \infty$ then every subsequence of $(a_n)$ is unbounded above:

Assume $a_n \to \infty$ and let $b_k = a_{n_k}$ be a subsequence on $(a_n)$. We must show that $(b_k)$ is unbounded above, that is, given any $K \in \mathbb{R}$, there exists some $k \in \mathbb{Z}^+$ such that $b_k > K$. So, let $K \in \mathbb{R}$ be given. Since $a_n \to \infty$, there exists $N \in \mathbb{Z}^+$ such that, for all $n \geq N$, $a_n > K$. Consider the sequence $(n_k)$ (defining the subsequence $b_k = a_{n_k}$). This is a strictly increasing sequence of positive integers, so $n_k \geq k$ for all $k$. Hence $n_N \geq N$, so $b_N = a_{n_N} > K$.

($\Leftarrow$) If every subsequence of $(a_n)$ is unbounded above then $a_n \to \infty$:

Instead of proving this directly, we will prove the contrapositive (recall that the *contrapositive* of an implication $P \Rightarrow Q$ is the statement $\neg Q \Rightarrow \neg P$ and that an implication and its contrapositive are logically equivalent; see section 7.2 for a reminder). That is, we will prove that if $a_n \not\to \infty$ then not every subsequence of $(a_n)$ is unbounded above. Equivalently, we will show that if $a_n \not\to \infty$ then there exists a subsequence of $(a_n)$ which is bounded above. Our first task is to understand precisely what the statement $a_n \not\to \infty$ means, and for this purpose the symbolic method developed in Chapter 7 is useful:

$$a_n \not\to \infty \iff \neg[\forall K \in \mathbb{R}, \exists N \in \mathbb{Z}^+, \forall n \in \mathbb{Z}^+ \cap [N, \infty), a_n > K]$$
$$\iff \exists K \in \mathbb{R}, \forall N \in \mathbb{Z}^+, \neg[\forall n \in \mathbb{Z}^+ \cap [N, \infty), a_n > K]$$
$$\iff \exists K \in \mathbb{R}, \forall N \in \mathbb{Z}^+, \exists n \in \mathbb{Z}^+ \cap [N, \infty), \neg[a_n > K]$$
$$\iff \exists K \in \mathbb{R}, \forall N \in \mathbb{Z}^+, \exists n \in \mathbb{Z}^+ \cap [N, \infty), a_n \leq K.$$

So, we assume that there exists some $K \in \mathbb{R}$ such that, for all $N \in \mathbb{Z}^+$, there is some $n \in \mathbb{Z}^+$ with $n \geq N$ such that $a_n \leq K$. From this assumption, we must show that $(a_n)$ has a subsequence which is bounded above.

The assumption holds for *all* $N \in \mathbb{Z}^+$, so in particular, it holds for $N = 1$. That is, there exists $n_1 \in \mathbb{Z}^+$ such that $a_{n_1} \leq K$. It also holds in the case $N = n_1 + 1$, so there exists $n_2 \in \mathbb{Z}^+$ such that $n_2 > n_1$ and $a_{n_2} \le K$. It also holds in the case $N = n_2 + 1$, so there exists $n_3 \in \mathbb{Z}^+$ such that $n_3 > n_2$ and $a_{n_3} \le K$, etc. Arguing in this way, we construct a strictly increasing sequence $(n_k)$ of positive integers such that for all $k$, $a_{n_k} \le K$. Hence, the subsequence $(a_{n_k})$ is bounded above (by $K$). $\qquad\square$

**Example 8.17.** $a_n = (1 + (-1)^n)n$ is unbounded above, but the subsequence $a_{2k+1} = 0$ is not, so by Proposition 8.16 $a_n \not\to \infty$. $\qquad\square$

**Corollary 8.18.** $a_n \to \infty$ *if and only if every subsequence of* $(a_n)$ *diverges to infinity.*

**Proof.** Since $(a_n)$ is a subsequence of itself, it is clear that if every subsequence of $(a_n)$ diverges to infinity, then $a_n \to \infty$.

Conversely, assume that $a_n \to \infty$. Assume, towards a contradiction, that there is a subsequence $(b_k) = (a_{n_k})$ which does *not* diverge to infinity. Then, by Proposition 8.16, $(b_k)$ has a subsequence $(c_m) = (b_{k_m})$ which is bounded above. But $(c_m)$ is a subsequence of $(a_n)$, so must be unbounded above (Proposition 8.16 again), a contradiction. $\qquad\square$

This corollary is rather analogous to one of the limit theorems we proved about convergent sequences (Theorem 4.3: if $a_n \to L$ then every subsequence of $(a_n)$ converges to $L$). In fact, many of the theorems we proved about convergent sequences have equivalents in this setting, and quite often the proofs are strikingly similar. For example, we have a version of the Squeeze Rule (Theorem 3.10) which one could call the "Sweep Rule".

**Proposition 8.19 (The Sweep Rule).** *If* $a_n \to \infty$ *and* $b_n \ge a_n$ *for all* $n$ *then* $b_n \to \infty$.

**Proof.** Let $K \in \mathbb{R}$ be given. Since $a_n \to \infty$, there exists $N \in \mathbb{Z}^+$ such that $a_n > K$ for all $n \ge N$. Hence, for all $n \ge N$, $b_n \ge a_n > K$. Hence $b_n \to \infty$. $\qquad\square$

We can also prove theorems about sequences which diverge to infinity which look rather similar to the Algebra of Limits (Theorem 3.5). Here's a couple of examples:

**Proposition 8.20.**

(i) If $a_n \to \infty$ and $b_n \to L$, then $a_n + b_n \to \infty$.

(ii) If $a_n \to \infty$ then $-a_n \to -\infty$.

*(iii) If $a_n \to -\infty$ then $-a_n \to \infty$.*

**Proof.**   (i) Let $K \in \mathbb{R}$ be given. We must show that there exists $N \in \mathbb{Z}^+$ such that, for all $n \geq N$, $a_n + b_n > K$.

Since $b_n \to L$, there exists $N_1 \in \mathbb{Z}^+$ such that, for all $n \geq N_1$, $|b_n - L| < 1$, and hence, $b_n > L - 1$. Let $K' = K - L + 1$. Since $a_n \to \infty$, there exists $N_2 \in \mathbb{Z}^+$ such that, for all $n \geq N_2$, $a_n > K'$. Let $N = \max\{N_1, N_2\}$. Then for all $n \geq N$,

$$\begin{aligned}
a_n + b_n &> a_n + L - 1 && \text{(since } n \geq N_1) \\
&> K' + L - 1 && \text{(since } n \geq N_2) \\
&= K
\end{aligned}$$

(ii) Let $K \in \mathbb{R}$ be given. We must show that there exists $N \in \mathbb{Z}^+$ such that, for all $n \geq N$, $-a_n < K$.

Now $a_n \to \infty$ and $-K \in \mathbb{R}$, so there exists $N \in \mathbb{Z}^+$ such that, for all $n \geq N$, $a_n > -K$. Hence, for all $n \geq N$, $-a_n < K$.

(iii) Exercise (just modify the proof of part (ii)).

$\square$

**Warning!** It's tempting to think that this Proposition is just a consequence of the Algebra of Limits (Theorem 3.5) and the algebraic rules $\infty + L = \infty$ and $(-1) \times \infty = -\infty$. But this is complete and utter *gibberish!* The symbol $\infty$ does not represent a real number, and it does not make sense to try to combine it with a real number (like $L$ or $-1$) using the algebraic operations of the field of real numbers (so $\infty + L$ and $(-1) \times \infty$ are meaningless strings of symbols). The point is that Definition 8.13 is *not* just a special case of Definition 2.5 with the real number $L$ replaced by the symbol $\infty$ or $-\infty$, so you can't directly deduce facts about sequences diverging to $\pm\infty$ from facts about convergent sequences.

**Corollary 8.21.** *$a_n \to -\infty$ if and only if every subsequence of $(a_n)$ is unbounded below.*

**Proof.** Exercise (it follows directly from Propositions 8.20 and 8.16).   $\square$

Returning to the subject of limits of *functions*, we can use sequences which diverge to infinity to define limits of functions at infinity. If a subset $D$ of $\mathbb{R}$ is unbounded above, it certainly contains sequences which diverge to $\infty$. (Proof: since $D$ is unbounded above, $n \in \mathbb{Z}^+$ is not an upper bound on $D$, so there exists $a_n \in D$ such that $a_n > n$. Now, as we saw in Example 8.14 $n \to \infty$, so $a_n \to \infty$ by the Sweep Rule.) Similarly, if $D$ is unbounded below then it contains sequences which diverge to $-\infty$. This allows us to make the following definition.

**Definition 8.22.** Let $f : D \to \mathbb{R}$ where $D$ is unbounded above. Then we say that $\lim_{x \to \infty} f(x) = L$ if, for all sequences $x_n \in D$ which diverge to $\infty$, $f(x_n) \to L$.

Similarly, let $f : D \to \mathbb{R}$ where $D$ is unbounded below. Then we say that $\lim_{x \to -\infty} f(x) = L$ if, given any sequence $x_n \in D$ which diverges to $-\infty$, $f(x_n) \to L$.

**Example 8.23.** Claim: $\lim_{x \to \infty} \dfrac{1}{x} = 0$.

**Proof.** Let $x_n \to \infty$. We must prove that $f(x_n) = 1/x_n \to 0$. So, let $\varepsilon > 0$ be given. Then there exists $N \in \mathbb{Z}^+$ such that $x_n > 1/\varepsilon$ for all $n \geq N$ (since $x_n \to \infty$). Hence, for all $n \geq N$,

$$0 < \frac{1}{x_n} < \varepsilon,$$

so $|f(x_n) - 0| < \varepsilon$. Hence $f(x_n) \to 0$. $\qquad\square$

**Example 8.24.** Claim: $\lim_{x \to -\infty} \dfrac{x+1}{x-1} = 1$

**Proof.** Let $x_n \to -\infty$. We must prove that $f(x_n) = (x_n+1)/(x_n-1) \to 1$. So, let $\varepsilon > 0$ be given. Then, since $x_n \to -\infty$, there exists $N \in \mathbb{Z}^+$ such that, for all $n \geq N$, $x_n < -2/\varepsilon$. Hence, for all $n \geq N$,

$$\begin{aligned}
|f(x_n) - 1| &= \left| \frac{x_n + 1}{x_n - 1} - 1 \right| = \frac{2}{|x_n - 1|} \\
&< \frac{2}{-x_n + 1} \qquad \text{(since } x_n < 1, \text{ so } x_n - 1 < 0\text{)} \\
&< \frac{2}{-x_n} < \varepsilon.
\end{aligned}$$

$\qquad\square$

Continuous functions $f : \mathbb{R} \to \mathbb{R}$ which have limits at both plus and minus infinity are, in many ways, analogous to continuous functions on closed bounded intervals $f : [a, b] \to \mathbb{R}$. In particular, we can prove something rather similar to the Extreme Value Theorem (Theorem 6.25).

**Theorem 8.25.** *Let $f : \mathbb{R} \to \mathbb{R}$ be continuous and have limits at plus and minus infinity. Then $f$ is bounded (above and below).*

**Proof.** It suffices to prove that every such function $f$ is bounded above, since $g = -f$ is also continuous with limits at plus and minus infinity, and if $g$ is bounded above then $f$ is bounded below.

Assume, towards a contradiction, that $f$ is unbounded above. Then, for each $n \in \mathbb{Z}^+$, there exists $x_n \in \mathbb{R}$ such that $f(x_n) > n$. Note that $f(x_n) \to \infty$ by the Sweep Rule, so every subsequence of $f(x_n)$ is unbounded above, by Proposition 8.16. Since $\lim_{x \to \infty} f(x)$ exists, $x_n \not\to \infty$ (if it did, $f(x_n)$ would be convergent, and hence bounded). Hence, by Proposition 8.16 there is a subsequence $(y_k) = (x_{n_k})$ of $(x_n)$ which is bounded above. Since $\lim_{x \to -\infty} f(x)$ exists, $y_k \not\to -\infty$ (if it did, $f(y_k) = f(x_{n_k})$ would be convergent, and hence bounded). Hence, by Corollary 8.21 there is a subsequence $(z_m) = (y_{k_m})$ of $(y_k)$ which is bounded below and above (since $(y_k)$ is bounded above). Hence, by the Bolzano–Weierstrass Theorem (Theorem 4.11), there is a subsequence $(w_p) = (z_{m_p})$ of $(z_m)$ which converges, $w_p \to a$ say. Note that $(w_p)$ is a subsequence of $(x_n)$. Since $f$ is continuous (at $a$), and $w_p \to a$, $f(w_p) \to f(a)$, and hence $f(w_p)$ is bounded. But $(f(w_p))$ is a subsequence of $(f(x_n))$ so must be unbounded above, a contradiction. $\square$

**Remark.** In the course of this proof we constructed a subsequence of a subsequence of a subsequence of a sequence!

**Example 8.26.** Consider the function

$$f : \mathbb{R} \to \mathbb{R}, \qquad f(x) = \frac{x^2 + 2|x|x}{x^2 + 1}.$$

This is continuous and has limits

$$\lim_{x \to \infty} f(x) = 3, \qquad \lim_{x \to \infty} f(x) = -1.$$

Hence, by Theorem 8.25, $f$ is bounded above and below. Its graph is depicted in Figure 8.1. Note that this function attains neither a maximum nor a minimum value. $\square$

In the case where $\lim_{x \to \infty} f(x) = \lim_{x \to -\infty} f(x)$, we can prove slightly more.

**Theorem 8.27.** *Let $f : \mathbb{R} \to \mathbb{R}$ be continuous and have equal limits at plus and minus infinity. Then $f$ is bounded (above and below) and $f$ attains a maximum or a minimum value (or both).*

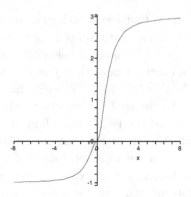

Fig. 8.1  A continuous, bounded function on $\mathbb{R}$ which attains neither a maximum nor a minimum.

**Proof.** By Theorem 8.25, $f$ is bounded above and below, so has a supremum $K$ say and an infimum $L$ say. If $K = L$, then the function is constant so clearly attains both a max and a min. So we may assume that $L < K$, in which case, at least one of them is not equal to the (common) limit $\lim_{x \to \pm\infty} f(x)$. We will consider the case $K \neq \lim_{x \to \pm\infty} f(x)$ and leave the case $L \neq \lim_{x \to \pm\infty} f(x)$ as an exercise.

Assume $K \neq \lim_{x \to \pm\infty} f(x)$. Since $K - \frac{1}{n} < K$, it is not an upper bound on $f$, so there exists $x_n \in \mathbb{R}$ such that $K - 1/n < f(x_n) \leq K$. By the Squeeze Rule, $f(x_n) \to K \neq \lim_{x \to \infty} f(x)$. Hence $x_n \not\to \infty$, so $(x_n)$ has a subsequence $y_k = x_{n_k}$ which is bounded above (Proposition 8.16). Now $f(y_k) \to K \neq \lim_{x \to -\infty} f(x)$, so $y_k \not\to -\infty$, and hence $(y_n)$ has a subsequence $z_m = y_{k_m}$ which is bounded below (Corollary 8.21) and above. By the Bolzano–Weierstrass Theorem $(z_m)$ has a convergent subsequence $(w_p)$, converging to $a \in \mathbb{R}$ say. Since $f$ is continuous (at $a$), $f(w_p) \to f(a)$. But $f(w_p)$ is a subsequence of $f(x_n)$, which converges to $K$, so $f(w_p) \to K$. Since the limit of a convergent sequence is unique, $f(a) = K$, that is, $f$ attains its supremum at $x = a$. Hence $f$ attains a maximum value. $\qquad \square$

**Example 8.28.** The function $f : \mathbb{R} \to \mathbb{R}$, $f(x) = \frac{x^2-1}{x^2+1}$ is continuous and has limits $\lim_{x \to \infty} f(x) = \lim_{x \to -\infty} f(x) = 1$. Hence, by Theorem 8.27, it is bounded and attains a max or a min (or both). In fact, this function attains a minimum (at $x = 0$) but does not attain a maximum value, as can be seen from its graph (Figure 8.2). $\qquad \square$

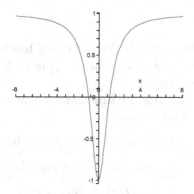

Fig. 8.2   The graph of $f(x) = (x^2 - 1)/(x^2 + 1)$.

## Hey! You now know what limits are!

It's worthwhile to pause and reflect on what we've achieved so far. We now have a completely precise and rigorous definition of $\lim_{x \to a} f(x)$ for functions $f : D \to \mathbb{R}$, where $D$ is a subset of $\mathbb{R}$. It follows that we now know, with no hand-waving or weasel words, precisely what the limit

$$f'(a) = \lim_{x \to a} \frac{f(x) - f(a)}{x - a}$$

means, that is, we have put the definition of the *derivative* of a function on a proper mathematical footing. Note that this is not a limit of $f : D \to \mathbb{R}$, but of the associated function

$$h : D \backslash \{a\} \to \mathbb{R}, \qquad h(x) = \frac{f(x) - f(a)}{x - a}.$$

The interesting, and somewhat curious, thing about this is that we managed to do this by studying *sequences* and, in particular, making precise what it means for sequences to converge. So, from the rather modest and unassuming definition of convergence of sequences (Definition 2.5) something very large and powerful has grown.

## 8.3   Summary

- Let $f : D \to \mathbb{R}$ where $D \subseteq \mathbb{R}$. We say that $f$ **has limit** $L$ **at** $a$, written $\lim_{x \to a} f(x) = L$, if for all sequences $(x_n)$ in $D \backslash \{a\}$ which converge to $a$, $f(x_n)$ converges to $L$.
- The **Algebra of Limits for functions** follows immediately from the Algebra of Limits for sequences:

$$\lim_{x \to a} (f(x) + g(x)) = \lim_{x \to a} f(x) + \lim_{x \to a} g(x),$$

$$\lim_{x \to a} f(x)g(x) = [\lim_{x \to a} f(x)][\lim_{x \to a} g(x)],$$

$$\lim_{x \to a} \frac{f(x)}{g(x)} = \frac{\lim_{x \to a} f(x)}{\lim_{x \to a} g(x)},$$

under suitable assumptions.
- A real sequence $(a_n)$ **diverges to infinity**, written $a_n \to \infty$, if for each $K \in \mathbb{R}$ there exists $N \in \mathbb{Z}^+$ such that for all $n \geq N$, $a_n > K$.
- Let $f : D \to \mathbb{R}$ where $D \subseteq \mathbb{R}$ is unbounded above. We say that $\lim_{x \to \infty} f(x) = L$ if for all sequences $(x_n)$ in $D$ such that $x_n \to \infty$, $f(x_n)$ converges to $L$.
- $a_n \to -\infty$ and $\lim_{x \to -\infty} f(x)$ are defined similarly.
- We can prove theorems about sequences which diverge to (plus or minus) infinity analogous to the limit theorems we proved for convergent sequences.

## 8.4 Tutorial problems

1. Show directly from Definition 8.1 that the following limit exists, and compute its value:
$$\lim_{x \to 2} \frac{x^3 - 8}{x^2 - 4}.$$

2. (a) Let $a_n \to L$ and $b_n \to -\infty$. Prove that $a_n + b_n \to -\infty$.

   (b) Let $a_n \to \infty$ and $b_n \to \infty$. Does it follow that $a_n - b_n \to 0$?

   (c) Let $a_n \to \infty$ and $b_n \to \infty$. Does it follow that $a_n/b_n \to 1$?

3. Let $f : \mathbb{R} \to \mathbb{R}$, $f(x) = (x^2 - 1)/(x^2 + 1)$. Prove directly from Definition 8.22 that $\lim_{x \to \infty} f(x) = 1$.

4. Let $f : \mathbb{R} \to \mathbb{R}$ be a function. Formulate a precise definition (using sequences) of the statement
$$\lim_{x \to \infty} f(x) = \infty.$$
Prove that $f(x) = x^3$ satisfies your definition.

## 8.5 Homework problems

1. Show directly from Definition 8.1 that the following limits exist, and compute their values.

   (a) $\displaystyle\lim_{x \to 2} \frac{x - 2}{x}$,  \qquad (b) $\displaystyle\lim_{x \to -1} \frac{x^3 + 1}{x^4 - 1}$,

   (c) $\displaystyle\lim_{h \to 0} \frac{(x + h)^4 - x^4}{h}$.

2. Determine whether each of the following statements is true. If true, prove it. If false, give a counterexample.

   (a) Let $a_n \to \infty$ and $b_n \to -\infty$. Then $a_n - b_n \to \infty$.

   (b) Let $a_n \to \infty$ and $b_n \to \infty$. Then $a_n b_n \to \infty$.

   (c) Let $a_n \to \infty$ and $b_n$ be unbounded above. Then $a_n + b_n \to \infty$.

   (d) Let $a_n \neq 0$ for all $n$, and $a_n \to 0$. Then $1/a_n \to \infty$.

3. Let $f : \mathbb{R}\backslash\{-1\} \to \mathbb{R}$, $f(x) = \dfrac{x + \sin x}{x + 1}$. Prove directly from Definition 8.22 that $\lim_{x \to \infty} f(x) = 1$.

4. Let $f : \mathbb{R}\backslash\{0\} \to \mathbb{R}$ be a function. Formulate a precise definition (using sequences) of the statement
$$\lim_{x \to 0} f(x) = \infty.$$
Prove that $f(x) = 1/x^2$ satisfies your definition.

# Chapter 9

# Differentiable functions

## 9.1 The main definition

Now that we have a rigorous definition of the limit of a function, it is straightforward to define derivatives.

**Definition 9.1.** Let $f : D \to \mathbb{R}$, where $D$ is some subset of $\mathbb{R}$, and $a \in D$ be a cluster point of $D$. Then $f$ is **differentiable at** $a$ if the limit

$$\lim_{x \to a} \frac{f(x) - f(a)}{x - a}$$

exists. In this case, we denote the limit $f'(a)$ and call it the **derivative of** $f$ **at** $a$. We say that $f$ is **differentiable** if it is differentiable at $a$ for all $a \in D$.

**Remarks**

- In general, in order to define $\lim_{x \to a} g(x)$, we only need $a$ to be a *cluster point* of the domain of $g$: it isn't necessary in general for $a$ to be in the domain of $g$, so $g(a)$ may or may not exist. For example

$$\lim_{x \to 1} \frac{x^2 - 1}{x - 1}$$

exists, although the function is undefined at $x = 1$. Note, however, that the limit defining $f'(a)$ contains the number $f(a)$, so to be differentiable at $a$, the point $a$ must be both an *element* and a *cluster point* of the domain of $f$. For example, a function $f : (0, \infty) \to \mathbb{R}$ cannot be differentiable at 0 (since $0 \notin (0, \infty)$), and a function $f : \mathbb{Z} \to \mathbb{R}$ cannot be differentiable anywhere (since $\mathbb{Z}$ has no cluster points).

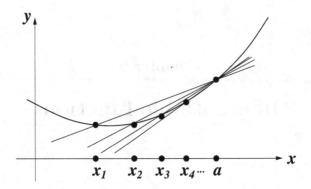

Fig. 9.1   A sequence of chords whose slopes converge to $f'(a)$.

- The limit in Definition 9.1 is defined in the usual way, using sequences (see Definition 8.1). That is, $f : D \to \mathbb{R}$ is differentiable at $a$ if, for every sequence $(x_n)$ in $D \backslash \{a\}$ such that $x_n \to a$, the sequence

$$s_n = \frac{f(x_n) - f(a)}{x_n - a}$$

converges. Geometrically, $s_n$ is the slope of the straight line passing through the points $(a, f(a))$ and $(x_n, f(x_n))$. Such a straight line is called a **chord** on the graph $y = f(x)$. As $x_n \to a$, the points defining the chord get "arbitrarily close" to one another (see Figure 9.1), and the chords approach a straight line through $(a, f(a))$ with slope $f'(a)$, the so-called **tangent line**. This is why the derivative $f'(a)$ is often interpreted as the slope of the graph $y = f(x)$ at the point $(a, f(a))$. Note, however, that this is just an *interpretation*. The *definition* of $f'(a)$ is in terms of limits which are, in turn, defined in terms of convergence of sequences.
- If $f : D \to \mathbb{R}$ is differentiable, then its derivative is also a function $f' : D \to \mathbb{R}$, mapping $D$ to $\mathbb{R}$. There is a very popular alternative notation for the derivative in this case, namely

$$f'(x) = \frac{df}{dx}.$$

This notation is convenient in some circumstances, but it tends to blur the distinction between a function (in this case $f'$) and the *value* of the function at a particular point (in this case $f'(x)$), so we will tend to avoid it.

**Example 9.2.** Let $f : \mathbb{R} \to \mathbb{R}$ such that $f(x) = c$, a constant. Then $f$ is differentiable (everywhere) and $f'(a) = 0$ for all $a \in \mathbb{R}$.

**Proof.** Let $(x_n)$ be any sequence in $\mathbb{R}\backslash\{a\}$ such that $x_n \to a$. Then

$$\frac{f(x_n) - f(a)}{x_n - a} = \frac{c - c}{x_n - a} = 0$$

which certainly converges, and its limit is 0. □

**Example 9.3.** Let $f : \mathbb{R} \to \mathbb{R}$ such that $f(x) = x$. Then $f$ is differentiable (everywhere) and $f'(a) = 1$ for all $a \in \mathbb{R}$.

**Proof.** Let $(x_n)$ be any sequence in $\mathbb{R}\backslash\{a\}$ such that $x_n \to a$. Then

$$\frac{f(x_n) - f(a)}{x_n - a} = \frac{x_n - a}{x_n - a} = 1$$

which certainly converges, and its limit is 1. □

**Example 9.4.** Let $f : \mathbb{R}\backslash\{0\} \to \mathbb{R}$ such that $f(x) = 1/x$. Then $f$ is differentiable (everywhere) and $f'(a) = -1/a^2$ for all $a \in \mathbb{R}\backslash\{0\}$.

**Proof.** Let $(x_n)$ be any sequence in $\mathbb{R}\backslash\{0, a\}$ such that $x_n \to a$. Then

$$\frac{f(x_n) - f(a)}{x_n - a} = \frac{1/x_n - 1/a}{x_n - a} = \frac{(a - x_n)/(ax_n)}{x_n - a} = -\frac{1}{ax_n} \to -\frac{1}{a^2}$$

by the Algebra of Limits (Theorem 3.5). □

**Example 9.5.** Let $f : \mathbb{R} \to \mathbb{R}$ such that $f(x) = |x|$. Then $f$ is differentiable at $a$ if and only if $a \neq 0$.

**Proof.** We establish this in three parts: if $a > 0$ then $f$ is differentiable at $a$; if $a < 0$ then $f$ is differentiable at $a$; and finally, $f$ is not differentiable at 0.

So, first assume that $a > 0$. Let $(x_n)$ be any sequence in $\mathbb{R}\backslash\{a\}$ such that $x_n \to a$. Then, by the definition of convergence, there exists $N \in \mathbb{Z}^+$ such that, for all $n \geq N$, $|x_n - a| < a/2$, and hence $x_n > a - a/2 > 0$. Hence, for all $n \geq N$,

$$s_n = \frac{f(x_n) - f(a)}{x_n - a} = \frac{x_n - a}{x_n - a} = 1$$

so the "tail" subsequence $s_{n+N}$ converges (it's constant!). Hence, by the Tail Lemma, $(s_n)$ converges, so $f$ is differentiable at $a$.

Second, assume that $a < 0$. Let $(x_n)$ be any sequence in $\mathbb{R}\backslash\{a\}$ such that $x_n \to a$. Then, by the definition of convergence, there exists $N \in \mathbb{Z}^+$

such that, for all $n \geq N$, $|x_n - a| < -a/2$, and hence $x_n < a - a/2 < 0$. Hence, for all $n \geq N$,

$$s_n = \frac{f(x_n) - f(a)}{x_n - a} = \frac{-x_n - (-a)}{x_n - a} = -1$$

so the "tail" subsequence $s_{n+N}$ converges. Hence, by the Tail Lemma, $(s_n)$ converges, so $f$ is differentiable at $a$.

Finally, let $x_n = (-1)^n/n$. Then $(x_n)$ is a sequence in $\mathbb{R}\backslash\{0\}$ converging to 0. But

$$s_n = \frac{f(x_n) - f(0)}{x_n - 0} = \frac{|(-1)^n/n|}{(-1)^n/n} = \frac{1}{(-1)^n} = (-1)^n$$

which does not converge. Hence $f$ is *not* differentiable at 0. □

**Remark.** To prove differentiability at $a$ we must consider *all* possible sequences in $D\backslash\{a\}$ which converge to $a$. It is not enough to prove that $(f(x_n) - f(a))/(x_n - a)$ converges for one particular sequence $(x_n)$. By contrast, to show that $f$ is *not* differentiable at $a$, it suffices to exhibit just one sequence $(x_n)$ such that $(f(x_n) - f(a))/(x_n - a)$ diverges.

There is a link between differentiablity and continuity, as we now show.

**Proposition 9.6.** *Let $f : D \to \mathbb{R}$ be differentiable at $a \in D$. Then $f$ is continuous at $a$.*

**Proof.** By Proposition 8.10, it suffices to show that $\lim_{x \to a} f(x) = f(a)$. Now, for all $x \in D\backslash\{a\}$,

$$f(x) - f(a) = g(x)(x - a), \qquad \text{where} \qquad g(x) = \frac{f(x) - f(a)}{x - a}.$$

By assumption, $g$ has a limit at $a$ (since $f$ is differentiable at $a$) and clearly $\lim_{x \to a}(x - a) = 0$. Hence, by the Algebra of Limits for functions (Theorem 8.12), $\lim_{x \to a}(f(x) - f(a)) = 0$, that is, $\lim_{x \to a} f(x) = f(a)$. □

**Remark.** The converse of Proposition 9.6 is false: if $f$ is continuous at $a$, a cluster point of its domain, it does not follow that $f$ is differentiable there. We've already seen a counteraxample: $f(x) = |x|$ is continuous at 0 but is not differentiable at 0 (see Example 9.5). In this case, the function fails to be differentiable only at a single isolated point. It is straightforward to construct functions which are *differentiable* only at a single isolated point, much as we constructed a function which was continuous only at a single isolated point in Example 6.7.

**Example 9.7.** Let $f : \mathbb{R} \to \mathbb{R}$ such that

$$f(x) = \begin{cases} 0, & x \in \mathbb{Q} \\ x^2, & x \notin \mathbb{Q} \end{cases}.$$

Then $f$ is differentiable at 0, and only at 0.

**Proof.** Let $a \in \mathbb{R}$ and assume that $f$ is differentiable at $a$. Then, by Propostion 9.6, $f$ is continuous at $a$. By Theorems 1.25 and 1.34, for each $n \in \mathbb{Z}^+$ there exist $r_n \in \mathbb{Q}$ and $i_n \in \mathbb{R} \backslash \mathbb{Q}$ such that $a < r_n < a + 1/n$ and $a < i_n < a + 1/n$. Clearly $r_n \to a$ and $i_n \to a$ by the Squeeze Rule. Hence, by the definition of continuity, $f(r_n) \to f(a)$ and $f(i_n) \to f(a)$. But $r_n$ is rational, so $f(r_n) = 0 \to 0$, and $i_n$ is irrational, so $f(i_n) = i_n^2 \to a^2$. Hence $a^2 = 0$, so $a = 0$. That is, $f$ is *not* differentiable at all $a \neq 0$.

It remains to show that $f$ *is* differentiable at 0. So, let $x_n \in \mathbb{R} \backslash \{0\}$ be any sequence converging to 0 and $s_n = (f(x_n) - f(0))/(x_n - 0) = f(x_n)/x_n$. We must show that $s_n$ converges (we will actually show it converges to 0). Let $\varepsilon > 0$ be given. Then, since $x_n \to 0$, there exists $N \in \mathbb{Z}^+$ such that, for all $n \geq N$, $|x_n - 0| < \varepsilon$. But then, for all $n \geq N$,

$$|s_n - 0| = |f(x_n)|/|x_n| = \begin{cases} 0, & \text{if } x_n \in \mathbb{Q} \\ |x_n|, & \text{if } x_n \notin \mathbb{Q} \end{cases} \leq |x_n| < \varepsilon.$$

Hence, $s_n \to 0$. □

## 9.2 The rules of differentiation

In principle, if we are given an explicit function, like $f : \mathbb{R} \to \mathbb{R}$, $f(x) = (x^3 + 1)^2$, we can show it is differentiable and compute its derivative by directly applying Definition 9.1. This quickly becomes complicated and tedious, however. So, much as we developed limit theorems to help us analyze convergence of complicated sequences by breaking them into simpler components (see Chapter 3), we next develop some "derivative theorems" which will allow us to do the same for differentiation. You are probably already familiar with these and are quite fluent in applying them. The point of this section is to show that, now we have a mathematically precise definition of the derivative, we can rigorously *prove* them, thus putting differential calculus on a solid foundation.

**Proposition 9.8 (Linearity).** *Let $f : D \to \mathbb{R}$, $g : D \to \mathbb{R}$ be differentiable at $a \in D$ with derivatives $f'(a)$ and $g'(a)$, respectively, and $c$ be a real constant. Then*

*(i) $cf : D \to \mathbb{R}$ is differentiable at $a$ with derivative $cf'(a)$.*
*(ii) $f + g : D \to \mathbb{R}$ is differentiable at $a$ with derivative $f'(a) + g'(a)$.*

**Proof.** Let $(x_n)$ be any sequence in $D\backslash\{a\}$ such that $x_n \to a$. Then

(i)
$$\frac{cf(x_n) - cf(a)}{x_n - a} = c\frac{f(x_n) - f(a)}{x_n - a} \to cf'(a)$$

(ii)
$$\frac{(f+g)(x_n) - (f+g)(a)}{x_n - a} = \frac{f(x_n) - f(a)}{x_n - a} + \frac{g(x_n) - g(a)}{x_n - a}$$
$$\to f'(a) + g'(a)$$

by the Algebra of Limits (Theorem 3.5).      □

**Proposition 9.9 (Product Rule).** *Let $f : D \to \mathbb{R}$, $g : D \to \mathbb{R}$ be differentiable at $a \in D$ with derivatives $f'(a)$ and $g'(a)$, respectively. Then $fg : D \to \mathbb{R}$ is differentiable at $a$ with derivative $f'(a)g(a) + f(a)g'(a)$.*

**Proof.** Let $(x_n)$ be any sequence in $D\backslash\{a\}$ such that $x_n \to a$. Then

$$\frac{f(x_n)g(x_n) - f(a)g(a)}{x_n - a} = \frac{(f(x_n) - f(a))g(x_n) + f(a)(g(x_n) - g(a))}{x_n - a}$$
$$= \frac{f(x_n) - f(a)}{x_n - a}g(x_n) + f(a)\frac{g(x_n) - g(a)}{x_n - a}.$$

Now $(f(x_n) - f(a))/(x_n - a) \to f'(a)$ (since $f$ is differentiable at $a$), and $(g(x_n) - g(a))/(x_n - a) \to g'(a)$ (since $g$ is differentiable at $a$). Further, by Proposition 9.6, $g$ is continuous at $a$, so $g(x_n) \to g(a)$. Hence, by the Algebra of Limits (Theorem 3.5),

$$\frac{f(x_n)g(x_n) - f(a)g(a)}{x_n - a} \to f'(a)g(a) + f(a)g'(a)$$

as was to be shown.      □

**Proposition 9.10.** *Every polynomial function $p : \mathbb{R} \to \mathbb{R}$, $p(x) = a_0 + a_1 x + \cdots + a_m x^m$ is differentiable, and its derivative is another polynomial function $p' : \mathbb{R} \to \mathbb{R}$, namely*

$$p'(x) = a_1 + 2a_2 x + \cdots + ma_m x^{m-1}.$$

**Proof.** Exercise: you can prove this by induction on the degree $m$ of the polynomial $p$, in much the same way that we proved Proposition 6.9.      □

The proofs of Propositions 9.8 and 9.9 were quite straightforward and analogous to the proofs of analagous parts of the Algebra of Limits (see Theorem 3.5). Our next differentiation rule, and its proof, are considerably more subtle.

**Theorem 9.11 (Chain Rule).** *Let* $f : A \to B$ *be differentiable at* $a \in A$ *and* $g : B \to \mathbb{R}$ *be differentiable at* $f(a) \in B$. *Then* $g \circ f : A \to \mathbb{R}$ *is differentiable at* $a$, *and*

$$(g \circ f)'(a) = g'(f(a))f'(a).$$

**Proof.** Let $(x_n)$ be any sequence in $A \backslash \{a\}$ such that $x_n \to a$ and

$$s_n = \frac{g(f(x_n)) - g(f(a))}{x_n - a}.$$

We must show that $s_n \to g'(f(a))f'(a)$. Consider the sequence $f(x_n)$ in $B$. Let

$$M = \{n \in \mathbb{Z}^+ : f(x_n) = f(a)\}$$
$$\overline{M} = \mathbb{Z}^+ \backslash M = \{n \in \mathbb{Z}^+ : f(x_n) \neq f(a)\}.$$

We will consider three cases separately: (i) $M$ is bounded above, (ii) $\overline{M}$ is bounded above, (iii) neither $M$ nor $\overline{M}$ is bounded above.

(i) *$M$ is bounded above*
In this case, there exists $N \in \mathbb{Z}^+$ such that, for all $n \geq N$, $f(x_n) \neq f(a)$. Let $y_n = x_{n+N}$. Then $y_n \to a$, and $f$ is continuous at $a$ (Proposition 9.6) so $f(y_n) \to f(a)$. Hence, $f(y_n)$ is a sequence in $B \backslash \{f(a)\}$ converging to $f(a)$, so

$$s_{n+N} = \frac{g(f(y_n)) - g(f(a))}{f(y_n) - f(a)} \frac{f(y_n) - f(a)}{y_n - a} \to g'(f(a)) \times f'(a)$$

by the Algebra of Limits (Theorem 3.5). Hence $s_n \to g'(f(a))f'(a)$ by the Tail Lemma.

(ii) *$\overline{M}$ is bounded above*
In this case, there exists $N \in \mathbb{Z}^+$ such that, for all $n \geq N$, $f(x_n) = f(a)$. Let $y_n = x_{n+N}$. Then $f(y_n) = f(a)$ for all $n$, so

$$s_{n+N} = \frac{g(f(y_n)) - g(f(a))}{y_n - a} = \frac{g(f(a)) - g(f(a))}{y_n - a} = 0 \to 0,$$

and hence $s_n \to 0$ by the Tail Lemma. But $(y_n)$ is a sequence in $A \backslash \{a\}$ converging to $a$, so, by the definition of derivative

$$0 = \frac{f(y_n) - f(a)}{y_n - a} \to f'(a),$$

whence $f'(a) = 0$. Hence $s_n \to g'(f(a))f'(a)$ as required.

(iii) *Neither $M$ nor $\overline{M}$ is bounded above*

Let $(m_k)$ be the increasing sequence taking all values in $M$ and $(\overline{m}_k)$ be the increasing sequence taking all values in $\overline{M}$. Then $s_{m_k}$ and $s_{\overline{m}_k}$ form a covering pair for $(s_n)$ so, by Lemma 4.8 it suffices to show that both these subseqences converge to $g'(f(a))f'(a)$. By definition, $f(x_{m_k}) = f(a)$ for all $k$, so

$$s_{m_k} = \frac{g(f(a)) - g(f(a))}{x_{m_k} - a} = 0 \to 0.$$

But $x_{m_k} \to a$ in $A \backslash \{a\}$, so $0 = (f(x_{m_k}) - f(a))/(x_{m_k} - a) \to f'(a)$, that is, $f'(a) = 0$. Hence, $s_{m_k} \to g'(f(a))f'(a)$ as required. Finally, $f(x_{\overline{m}_k}) \neq f(a)$ for all $k$, so

$$s_{\overline{m}_k} == \frac{g(f(x_{\overline{m}_k})) - g(f(a))}{f(x_{\overline{m}_k}) - f(a)} \frac{f(x_{\overline{m}_k}) - f(a)}{x_{\overline{m}_k} - a} \to g'(f(a)) \times f'(a)$$

by the Algebra of Limits (Theorem 3.5). $\qquad\square$

**Example 9.12.** Let $f : \mathbb{R} \to \mathbb{R}$ such that $f(x) = (x^3 + 1)^2$. Then $f$ is differentiable and $f'(x) = 6(x^3 + 1)x^2$.

**Proof.** $f = h \circ g$ where $g(x) = x^3 + 1$ and $h(x) = x^2$. Now $g, h$ are differentiable, with derivatives $g'(x) = 3x^2$ and $h'(x) = 2x$ (by Propositions 9.8 and 9.10). Hence, by the Chain Rule, $f$ is differentiable, and

$$f'(x) = h'(g(x))g'(x) = 2g(x)3x^2 = 6(x^3 + 1)x^2.$$

$\qquad\square$

**Example 9.13.** Let $m$ be a positive integer and $f : \mathbb{R} \backslash \{0\}$ be the function $f(x) = 1/x^m$. Then $f$ is differentiable, and $f'(x) = -m/x^{m+1}$.

**Proof.** $f = h \circ g$ where $g(x) = x^m$ and $h(x) = 1/x$, both of which are differentiable. Hence, by the Chain Rule, so is $f$, and

$$f'(x) = h'(g(x))g'(x) = \frac{-1}{g(x)^2}mx^{m-1} = -\frac{m}{x^{m+1}}$$

by Example 9.4 and Proposition 9.10. $\qquad\square$

Generalizing the trick used in this proof, we obtain another useful rule of differentiation:

**Corollary 9.14 (Quotient Rule).** *Let* $f : D \to \mathbb{R}$ *and* $g : D \to \mathbb{R}$, *be differentiable at* $a \in D$ *and* $g(a) \neq 0$. *Then* $f/g$ *is differentiable at* $a$ *and*

$$\left(\frac{f}{g}\right)'(a) = \frac{f'(a)g(a) - f(a)g'(a)}{g(a)^2}.$$

**Proof.** Let $h(x) = 1/g(x) = r(g(x))$, where $r : \mathbb{R}\backslash\{0\} \to 0$ is the function $r(x) = 1/x$. By the Chain Rule and Example 9.4, $h$ is differentiable at $a$ with derivative

$$h'(a) = r'(g(a))g'(a) = \frac{-1}{g(a)^2}g'(a).$$

Hence, by the Product Rule, $f/g = fh$ is differentiable at $a$ with derivative

$$(f/g)'(a) = f'(a)h(a) + f(a)h'(a) = \frac{f'(a)}{g(a)} - \frac{f(a)g'(a)}{g(a)^2}$$

$$= \frac{f'(a)g(a) - f(a)g'(a)}{g(a)^2}.$$

as was claimed. $\qquad\square$

So we now know that the function $f(x) = x^k$ is differentiable for all integers $k$ (everywhere it is well-defined, i.e. everywhere except 0 if $k < 0$), and that its derivative is $f'(x) = kx^{k-1}$. Recall that we can also define $f : (0, \infty) \to \mathbb{R}$, $f(x) = x^r$ where $r$ is any *rational* number: if $r = p/q$, we define $x^r = (x^{1/q})^p$. It turns out that this function is also differentiable, with the expected derivative:

**Proposition 9.15.** *Let* $r$ *be a rational number and* $f : (0, \infty) \to \mathbb{R}$ *be the function* $f(x) = x^r$. *Then* $f$ *is differentiable and*

$$f'(x) = rx^{r-1}.$$

**Proof.** This is a special case of Proposition 12.11, which will be formulated and proved in Chapter 12. $\qquad\square$

It may occur to you that one could prove Proposition 9.15 by arguing as follows. Let $r = p/q$ where $p, q \in \mathbb{Z}^+$, and define $g(x) = f(x)^q = x^p$. Then this is certainly differentiable, with derivative $px^{p-1}$. Hence, by the Chain Rule,

$$px^{p-1} = qf(x)^{q-1}f'(x)$$

$$\Rightarrow \quad f'(x) = \frac{px^{p-1}}{q(x^{p/q})^{q-1}} = \frac{p}{q}x^{p-1-\frac{p(q-1)}{q}}$$

$$= \frac{p}{q}x^{\frac{p}{q}-1} = rx^{r-1}.$$

The only thing wrong with this "proof" is that, in applying the Chain Rule (to $F = h \circ f$ where $h(x) = x^q$) you are presupposing that $f$ is differentiable, that is, you're assuming what you want to prove! So all this argument really shows is that *if* $f$ is differentiable, then its derivative must be $f'(x) = rx^{r-1}$.

## 9.3 Functions differentiable on an interval

Two applications of differential calculus are particularly pervasive in the natural sciences. In one, we use the derivative of a function to deduce its maximum (or minimum) value and the input which produces that value. In the second, we are given information, not about the function itself, but about its derivative (which is often interpreted as a "rate of change" with respect to time). We then seek to reconstruct the function itself from this information. These two applications rely fundamentally on the Interior Extremum Theorem, and the Mean Value Theorem, respectively, and the purpose of this section is to state and prove these. Both are statements concerning functions which are differentiable (at least) on an open interval $(a, b)$. The functions in question may be differentiable elsewhere too (perhaps on the whole of $\mathbb{R}$, in fact), so it's important when you read "$f$ is differentiable on $(a, b)$" not to add the unwarranted assumption "and only on $(a, b)$".

Recall that $f : D \to \mathbb{R}$ attains a **maximum** at $a \in D$ if $f(x) \leq f(a)$ for all $x \in D$. Similarly, $f$ attains a **minimum** at $a \in D$ if $f(x) \geq f(a)$ for all $x \in D$. We say that $f$ attains an **extremum** at $a \in D$ if it attains either a maximum or minimum at $a$. (The word "extremum" means "maximum or minimum", in much the same way that "monotonic" means "increasing or decreasing". Its plural is "extrema".)

**Theorem 9.16 (The Interior Extremum Theorem).** *Let* $f : (a, b) \to \mathbb{R}$ *be differentiable and $f$ attain an extremum at* $c \in (a, b)$. *Then* $f'(c) = 0$.

**Proof.** Without loss of generality, we may assume that $f$ attains a maximum at $c$, since if $f$ attains a minimum at $c$ then $g = -f$ is also differentiable on $(a, b)$ and attains a maximum at $c$, and $g'(c) = 0$ implies $f'(c) = 0$.

So, assume $f$ attains a maximum at $c \in (a, b)$. By assumption, $f'(c)$ exists. Let $x_n = c + (b - c)/(n + 1) \in (c, b)$. Then $x_n \to c$, so

$$\frac{f(x_n) - f(c)}{x_n - c} \to f'(c).$$

But $f(c)$ is a maximum of $f$, so $f(x_n) - f(c) \le 0$, and $x_n - c > 0$, so all terms in the above sequence are non-positive. Hence, by Proposition 3.4, its limit is non-positive, i.e. $f'(c) \le 0$.

Now let $y_n = c + (c - a)/(n + 1) \in (a, c)$. Then $y_n \to c$, so

$$\frac{f(y_n) - f(c)}{x_n - c} \to f'(c).$$

But $f(c)$ is a maximum of $f$, so $f(y_n) - f(c) \le 0$, and $y_n - c < 0$, so all terms in the above sequence are non-negative. Hence, by Proposition 3.4, its limit is non-negative, i.e. $f'(c) \ge 0$.

Hence, $f'(c) = 0$, as was to be shown. □

**Definition 9.17.** Let $f : D \to \mathbb{R}$ be differentiable. Then $c \in D$ is a **critical point** of $f$ if $f'(c) = 0$.

**Example 9.18.** Find the maximum and minimum values attained by the function

$$f : [0, 3] \to \mathbb{R}, \qquad f(x) = x^2 - 2x + 7.$$

*Solution:* First note that, since $f$ is continuous, it certainly attains both a maximum and a minimum by the Extreme Value Theorem (Corollary 6.25). Since $f$ is differentiable on $(0, 3)$, it follows from the Interior Extremum Theorem that each extremum occurs either at an endpoint, that is, 0 or 3, or at an interior critical point of $f$. Now

$$f'(x) = 2x - 2 = 0$$

if and only if $x = 1$, so 1 is the only critical point of $f$. Since $f(1) = 6$, $f(0) = 7$, and $f(3) = 10$, we deduce that $f$ attains a maximum value of 10 at the right endpoint 3, and a minimum value of 6 at the interior critical point 1.

In this case, using calculus was really overkill, since we could have deduced the same information by simply completing the square:

$$f(x) = (x - 1)^2 + 6.$$

□

The next theorem has hypotheses (assumptions on $f$) which look very restrictive – so restrictive that one might wonder whether the theorem is of any practical use at all. In fact, as we shall see, it has very powerful and useful consequences.

**Theorem 9.19 (Rolle's Theorem).** *Let $f : [a, b] \to \mathbb{R}$ be continuous and differentiable on $(a, b)$. Assume $f(a) = f(b)$. Then there exists $c \in (a, b)$ such that $f'(c) = 0$.*

**Proof.** By the Extreme Value Theorem, $f$ attains both a maximum and a minimum value on $[a, b]$. If both of these occur at the endpoints of $[a, b]$ then, since $f(a) = f(b)$, the maximum value coincides with the minimum value, whence it follows that $f : [a, b] \to \mathbb{R}$ is constant. But then $f'(x) = 0$ for all $x \in (a, b)$, so $c = (b - a)/2$, for example, has $f'(c) = 0$.

Hence, we may assume that either the maximum or the minimum value does not occur at an endpoint. But then $f$ attains an extremum at some interior point $c \in (a, b)$, so $f'(c) = 0$ by the Interior Extremum Theorem. $\square$

**Theorem 9.20 (The Mean Value Theorem).** *Let $f : [a, b] \to \mathbb{R}$ be continuous and differentiable on $(a, b)$. Then there exists $c \in (a, b)$ such that*

$$f'(c) = \frac{f(b) - f(a)}{b - a}.$$

**Proof.** Let $g : [a, b] \to \mathbb{R}$ such that

$$g(x) = f(x) - \frac{f(b) - f(a)}{b - a}(x - a).$$

Then $g$ is continuous, and is differentiable on $(a, b)$. Furthermore $g(a) = f(a)$ and $g(b) = f(a)$. So $g$ satisfies the hypotheses of Rolle's Theorem, and we deduce that there exists $c \in (a, b)$ such that $g'(c) = 0$. But

$$g'(c) = f'(c) - \frac{f(b) - f(a)}{b - a}$$

which completes the proof. $\square$

**Remarks**

- The Mean Value Theorem has a nice geometric interpretation. Consider the graph $y = f(x)$ of a differentiable function $f : \mathbb{R} \to \mathbb{R}$. For each pair of distinct numbers $a < b$, we can construct the chord (straight line) passing through the points $(a, f(a))$ and $(b, f(b))$ on the graph. Its slope is $(f(b) - f(a))/(b - a)$. The Mean Value Theorem asserts that, at some point $(c, f(c))$ on the graph between these two points, the tangent line to the graph is parallel to the chord (see Figure 9.2).

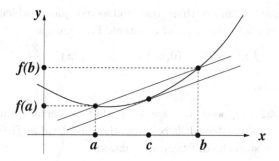

Fig. 9.2 Geometric interpretation of the Mean Value Theorem.

- The Mean Value Theorem guarantees the existence of a point $c$ with the required $f'(c)$. It says nothing about uniqueness: there could be more than one such point.
- Clearly Rolle's Theorem is a special case of the Mean Value Theorem (geometrically, the case where the chord is horizontal). It's slightly surprising at first sight, therefore, that Rolle's Theorem actually implies the Mean Value Theorem.

You perhaps don't realize it, but you have probably been using the Mean Value Theorem rather a lot. We have seen (Example 9.2) that if a function is constant then its derivative is zero everywhere. When one solves a differential equation, one often uses the converse fact: if a function has zero derivative everywhere, then it must be constant. But how do we *know* this? It's a consequence of the Mean Value Theorem!

**Corollary 9.21.** *Let* $I \subseteq \mathbb{R}$ *be an interval and* $f : I \to \mathbb{R}$ *have derivative* $f'(x) = 0$ *for all* $x \in I$. *Then* $f$ *is constant.*

**Proof.** Assume, to the contrary, that there exist points $a, b \in I$, with $a < b$, such that $f(a) \neq f(b)$. Then $f$ is continuous on $[a, b]$ and differentiable on $(a, b)$, so by the Mean Value Theorem, there exists $c \in (a, b)$ such that
$$f'(c) = \frac{f(b) - f(a)}{b - a} \neq 0.$$
But $f'(x) = 0$ for all $x \in I$, a contradiction. $\qquad\square$

**Warning!** Corollary 9.21 concerns functions on a single interval whose derivatives are everywhere zero. It's easy to write down functions whose domains consist of more than one connected piece which have zero derivative everywhere but are *not* constant. For example

$$f : (-\infty, 0) \cup (0, \infty) \to \mathbb{R}, \qquad f(x) = \frac{x}{|x|}$$

has this property.

**Example 9.22.** Suppose we are told that a certain function $f$ is differentiable on the interval $[0, \infty)$, has the initial value $f(0) = -1$, and for all $x \geq 0$ satisfies the differential equation

$$f'(x) = xf(x)^2.$$

Does this information uniquely determine $f$? If so, can we find a formula for it? Yes, and yes:

First, note that, since $f(0) = -1 \neq 0$ and $f$ is differentiable, hence continuous, there exists $b > 0$ such that $f(x) \neq 0$ for all $x \in [0, b]$. (If not, there would be a sequence $x_n > 0$, $x_n \to 0$, such that $f(x_n) = 0$, which, since $f$ is continuous, would imply that $f(0) = \lim f(x_n) = 0$.) Now, on $[0, b]$ we may rearrange the differential equation as

$$\frac{d}{dx}\left(\frac{-1}{f(x)} - \frac{x^2}{2}\right) = 0.$$

Since the expression in brackets has zero derivative on the interval $[0, b]$, it follows from Corollary 9.21 that it is constant, that is,

$$\frac{-1}{f(x)} - \frac{x^2}{2} = \frac{-1}{f(0)} - \frac{0^2}{2} = 1.$$

Hence, for all $x \in [0, b]$,

$$f(x) = \frac{-2}{x^2 + 2}.$$

But $b > 0$ was, by definition, any number such that $f(x) \neq 0$ for all $x \in [0, b]$. Clearly, the above expression never vanishes, so this equation for $f$ holds on $[0, b]$ for all $b > 0$. That is,

$$f : [0, \infty) \to \mathbb{R}, \qquad f(x) = \frac{-2}{x^2 + 2}$$

is the unique solution of the differential equation with $f(0) = -1$. $\quad\square$

When we first learn calculus, we are taught to think of $f'(x)$ as the "rate of change of $f$ with respect to the variable $x$". So if, at a particular point $a \in D$, $f'(a) > 0$, we expect that $f$ should be strictly increasing (in the sense of Definition 6.15), at least sufficiently close to $a$. To state this more precisely, one might expect that if $f'(a) > 0$ then there exists $\varepsilon > 0$ such that the function $f : (a - \varepsilon, a + \varepsilon) \to \mathbb{R}$ is strictly increasing. This is what we called "Assertion 2" in the Foreword to this book. In fact, this naive expectation is **false**, as the next example demonstrates.

**Example 9.23.** Let $f : \mathbb{R} \to \mathbb{R}$ be defined such that

$$f(x) = \begin{cases} x & \text{if } x \in \mathbb{Q}, \\ x - x^2 & \text{if } x \notin \mathbb{Q}. \end{cases}$$

I claim that $f$ is differentiable at 0, that $f'(0) = 1 > 0$, but that $f$ is not increasing on any open interval containing 0.

**Proof.** We first show that $f$ is differentiable at 0 with $f'(0) = 1$. So, let $(x_n)$ be any sequence in $\mathbb{R}\backslash\{0\}$ converging to 0, and consider

$$s_n = \frac{f(x_n) - f(0)}{x_n - 0} = \frac{f(x_n)}{x_n}.$$

If $x_n$ is rational, then $|s_n - 1| = |x_n/x_n - 1| = 0$, whereas if $x_n$ is irrational then

$$|s_n - 1| = \left| \frac{x_n - x_n^2}{x_n} - 1 \right| = |x_n|.$$

Hence, for all $n$, $0 \le |s_n - 1| \le |x_n|$, so $|s_n - 1| \to 0$ by the Squeeze Rule. Hence $s_n \to 1$. It follows that $f'(0) = 1$, as claimed.

We now show that $f$ is not increasing on any open interval containing 0. So, let $\varepsilon > 0$ be given. Then, by Theorem 1.34, there exists an irrational number $x$ such that $0 < x < \varepsilon$. By definition, $f(x) = x - x^2 < x$. By Theorem 1.25, there exists a *rational* number $y$ such that $x - x^2 < y < x$. But then $y < x$ and $f(y) = y > x - x^2 = f(x)$. Hence, $f$ is not increasing on the interval $(-\varepsilon, \varepsilon)$. This is true no matter which positive number $\varepsilon$ we choose, which establishes the claim. $\qquad\square$

The moral of the story is that just knowing the sign of $f'$ at a single isolated point gives us very little useful information about the behaviour of the function. If, on the other hand, we know that $f'$ is positive everywhere on some interval, then it *does* follow that $f$ is strictly increasing on that interval. This is another useful consequence of the Mean Value Theorem.

**Proposition 9.24.** *Let $I \subseteq \mathbb{R}$ be an interval and $f : I \to \mathbb{R}$ be differentiable. If $f'(x) > 0$ for all $x \in I$, then $f$ is strictly increasing, and hence is injective.*

**Proof.** Assume, to the contrary, that there exist $a, b \in I$ with $a < b$ such that $f(a) \geq f(b)$. Then, by the Mean Value Theorem, there exists $c \in (a, b)$ such that

$$f'(c) = \frac{f(b) - f(a)}{b - a} \leq 0.$$

But this contradicts the assumption that $f'(x) > 0$ for all $x \in I$. It follows that $f$ is injective (Lemma 6.16). $\qquad \square$

**Remarks**

- We can easily modify the proof to show that if $f'(x) < 0$ for all $x \in I$, then $f$ is strictly *decreasing*.
- Proposition 9.24 gives us a sneaky way of showing that some functions are injective.

**Example 9.25.** Let $f : \mathbb{R} \to \mathbb{R}$ such that $f(x) = 2x + \sin x$. Then $f$ is strictly increasing, and hence injective.

**Proof.** The function $f$ is differentiable, with derivative[1]

$$f'(x) = 2 + \cos x \geq 1 > 0.$$

Hence, by Proposition 9.24, $f$ is strictly increasing, and hence injective. (It's an interesting exercise to try to prove that $f$ is injective directly from the definition of injectivity, without using calculus.) $\qquad \square$

It's important to realize that Proposition 9.24 applies only to functions defined on an *interval*. In general, a differentiable function $f : D \to \mathbb{R}$ can have positive derivative everywhere but still fail to be increasing.

**Example 9.26.** Let $f : \mathbb{R} \backslash \{0\} \to \mathbb{R}$ such that $f(x) = -\frac{1}{x}$. Then

$$f'(x) = \frac{1}{x^2} > 0$$

for all $x \in \mathbb{R} \backslash \{0\}$, but $f$ is not increasing, since, for example, $f(-1) > f(1)$.

Sometimes a more general version of the Mean Value Theorem, concerning a *pair* of differentiable functions on an interval, is useful.

---

[1]Strictly speaking, you haven't yet proved that sin is differentiable, with derivative cos, so this argument requires you to suspend disbelief until Chapter 10.

**Theorem 9.27 (Extended Mean Value Theorem).** *Let $f$ and $g$ be real functions which are continuous on $[a, b]$ and differentiable on $(a, b)$, and assume that, for all $x \in (a, b)$, $g'(x) \neq 0$. Then there exists $c \in (a, b)$ such that*

$$\frac{f'(c)}{g'(c)} = \frac{f(b) - f(a)}{g(b) - g(a)}.$$

**Proof.** First note that the conditions on $g$ imply that $g(a) \neq g(b)$ (since if $g(b) = g(a)$ then $g'(x) = 0$ for some $x \in (a, b)$ by Rolle's Theorem). Hence,

$$\alpha = \frac{f(b) - f(a)}{g(b) - g(a)}$$

is a well-defined real constant. Let $h : [a, b] \to \mathbb{R}$ such that $h(x) = f(x) - \alpha g(x)$. Then $h$ is continuous on $[a, b]$ and differentiable on $(a, b)$, and, as one may easily verify,

$$h(a) = \frac{f(a)g(b) - g(a)f(b)}{g(b) - g(a)} = h(b).$$

Hence $h$ satisfies the hypotheses of Rolle's Theorem, and we conclude that there exists $c \in (a, b)$ such that $h'(c) = 0$. But then $f'(c) = \alpha g'(c)$, whence the result immediately follows. $\square$

If we apply Theorem 9.27 in the case where $g(x) = x$, we immediately deduce the usual Mean Value Theorem. We will see that other choices of $g$ can be useful. For example, we can use Theorem 9.27 to rigorously justify a popular trick for computing limits called "L'Hospital's Rule".

**Theorem 9.28 (L'Hospital's Rule).** *Let $I \subseteq \mathbb{R}$ be an open interval, $f, g : I \to \mathbb{R}$ be differentiable functions, $a \in I$, and $f(a) = g(a) = 0$. Assume that, for all $x \in I$, $g'(x) \neq 0$ and all $x \in I \setminus \{a\}$, $g(x) \neq 0$. Then*

$$\lim_{x \to a} \frac{f(x)}{g(x)} = \lim_{x \to a} \frac{f'(x)}{g'(x)}$$

*provided the limit on the right exists.*

**Proof.** Let $(x_n)$ be any sequence in $I \setminus \{a\}$ converging to $a$, and for each $n$, consider $f, g$ restricted to the closed interval with endpoints $a$ and $x_n$. These functions satisfy the hypotheses of Theorem 9.27, so there exists $c_n$, between $a$ and $x_n$, such that

$$\frac{f'(c_n)}{g'(c_n)} = \frac{f(x_n) - f(a)}{g(x_n) - g(a)} = \frac{f(x_n)}{g(x_n)}.$$

By the Squeeze Rule, $c_n \to a$, and so, assuming the limit

$$\lim_{x \to a} \frac{f'(x)}{g'(x)}$$

exists, call it $L \in \mathbb{R}$, say, $f'(c_n)/g'(c_n) \to L$ (by the definition of limit). Hence $f(x_n)/g(x_n) \to L$, as was to be shown. □

**Example 9.29.** Compute $\lim\limits_{x \to 0} \dfrac{1 - \cos x}{\sin^2 x}$.

*Solution:*　　Let $I = (-\pi/2, \pi/2)$, $f(x) = 1 - \cos x$, $g(x) = \sin^2 x$, and $a = 0 \in I$. Then

$$\lim_{x \to 0} \frac{f'(x)}{g'(x)} = \lim_{x \to 0} \frac{\sin x}{2 \sin x \cos x} = \frac{1}{2}.$$

Hence, by L'Hospital's Rule,

$$\lim_{x \to 0} \frac{f(x)}{g(x)} = \frac{1}{2}$$

also. □

## 9.4　Higher derivatives and Taylor's Theorem

If a function $f : D \to \mathbb{R}$ is differentiable (everywhere on $D$) then its derivative defines another function $f' : D \to \mathbb{R}$, and we can ask whether this function, in turn, is differentiable. If it is, then its derivative is denoted $f'' : D \to \mathbb{R}$ and called the *second derivative* of $f$. Similarly, if $f'' : D \to \mathbb{R}$ is differentiable, its derivative $f''' : D \to \mathbb{R}$ is the *third derivative* of $f$. Proceeding inductively, we can define the $n^{\text{th}}$ derivative of $f$, denoted $f^{(n)}$ to be the derivative (if it exists), of $f^{(n-1)}$, where $f^{(0)}$ is, by definition, just the function $f$ itself (so $f^{(1)} = f'$, $f^{(2)} = f''$ etc.). If $f^{(n)} : D \to \mathbb{R}$ exists, we say that $f$ is $n$ times differentiable, and if $f$ is $n$ times differentiable for all $n \in \mathbb{Z}^+$, we say that $f$ is *smooth*.

**Example 9.30.** Every polynomial function $p : \mathbb{R} \to \mathbb{R}$, $p(x) = a_0 + a_1 x + \cdots + a_m x^m$, is smooth. This follows immediately from Proposition 9.10.

**Example 9.31.** The function $f : \mathbb{R} \to \mathbb{R}$, $f(x) = |x|^3$ is twice differentiable but not three times differentiable. To see this, note that $f$ coincides with the smooth function $f_+(x) = x^3$ on $(0, \infty)$ and with the smooth function

$f_-(x) = -x^3$ on $(-\infty, 0)$, so it suffices to consider the differentiability of $f$ at $0$. Let $x_n$ be a sequence in $\mathbb{R}\backslash 0$ converging to $0$. Then

$$\left| \frac{f(x_n) - f(0)}{x_n - 0} \right| = |x_n|^2 \to 0$$

so $f'(0) = 0$, and it follows that $f$ is once differentiable. In fact

$$f'(x) = \begin{cases} f'_+(x) = 3x^2 & x > 0 \\ 0 & x = 0 \\ f'_-(x) = -3x^2 & x < 0 \end{cases},$$

so

$$\left| \frac{f'(x_n) - f'(0)}{x_n - 0} \right| = |3x_n| \to 0,$$

whence $f''(0) = 0$. Hence, $f$ is twice differentiable, and

$$f''(x) = \begin{cases} f''_+(x) = 6x & x > 0 \\ 0 & x = 0 \\ f''_-(x) = -6x & x < 0, \end{cases} = 6|x|.$$

But $f''$ is not differentiable at $0$, by Example 9.5, so $f$ is not three times differentiable.

An interesting interpretation of the Mean Value Theorem can be given as follows. Let $I \subseteq \mathbb{R}$ be an open interval, $f : I \to \mathbb{R}$ be differentiable, and choose and fix some $a \in I$. Then for any other $x \in I$, we may apply the Mean Value Theorem to $f : [a, x] \to \mathbb{R}$ (if $x > a$) or $f : [x, a] \to \mathbb{R}$ (if $x < a$) to deduce that there exists $c$ between $a$ and $x$ such that

$$f(x) = f(a) + f'(c)(x - a).$$

Note that the same equation holds trivially in the case where $x = a$ (with $c = x = a$). So, for $x$ close to $a$ (meaning that $|x - a|$ is small) we can approximate $f$ by the constant $f(a)$, and the error in this approximation is of size $|f'(c)||x - a|$, where $c$ is somewhere between $a$ and $x$ (and, of course, depends on $x$).

The moral is that differentiability (once) of $f$ on an open interval allows one to approximate $f$ by a degree $0$ polynomial function (i.e. a constant), with an error controlled by $f'$. It turns out that if $f$ is $n$ times differentiable one can do better: one can approximate $f$ by a degree $(n-1)$ polynomial function, with an error controlled by $f^{(n)}$.

**Theorem 9.32 (Taylor's Theorem).** *Let $I \subseteq \mathbb{R}$ be an open interval, $f : I \to \mathbb{R}$ be $(n+1)$ times differentiable, and $a, x \in I$. Then there exists $c$ between $a$ and $x$ such that*

$$f(x) = f(a) + f'(a)(x-a) + \frac{f''(a)}{2!}(x-a)^2 + \cdots + \frac{f^{(n)}(a)}{n!}(x-a)^n$$
$$+ \frac{f^{(n+1)}(c)}{(n+1)!}(x-a)^{n+1}.$$

**Proof.** First note that the claimed equation holds trivially (with $c = a$) in the case where $x = a$. Choose and fix $a, x \in I$, $x \neq a$, and consider the function $F : I \to \mathbb{R}$,

$$F(t) = f(x) - f(t) - f'(t)(x-t) - \frac{f''(t)}{2!}(x-t)^2 - \cdots - \frac{f^{(n)}(t)}{n!}(x-t)^n.$$

Then the claim to be proved is that there exists $c$ between $a$ and $x$ such that

$$F(a) = \frac{f^{(n+1)}(c)}{(n+1)!}(x-a)^{n+1}.$$

Now, by the Product Rule, $F$ is differentiable, and, as is easily verified,

$$F'(t) = -\frac{f^{(n+1)}(t)}{n!}(x-t)^n.$$

Let $G(t) = (x-t)^{n+1}$. Then $F, G$ satisfy the hypotheses of Theorem 9.27 on the interval with endpoints $a, x$ (note that $G'(t) = 0$ only if $t = x$), so there exists $c$ between $a$ and $x$ such that

$$\frac{F'(c)}{G'(c)} = \frac{F(x) - F(a)}{G(x) - G(a)} = \frac{F(a)}{G(a)}$$

i.e. $$\frac{-f^{(n+1)}(c)(x-c)^n}{n![-(n+1)(x-c)^n]} = \frac{F(a)}{(x-a)^{n+1}},$$

whence the claimed expression for $F(a)$ immediately follows. $\qquad\square$

The polynomial

$$p_n(x) = f(a) + f'(a)(x-a) + \frac{f''(a)}{2!}(x-a)^2 + \cdots + \frac{f^{(n)}(a)}{n!}(x-a)^n$$

is often called the $n^{\text{th}}$ **Taylor approximant** of the function $f$ about $a$, and

$$\frac{f^{(n+1)}(c)}{(n+1)!}(x-a)^{n+1}$$

is called the **remainder**.

**Example 9.33.** Let $f : [0, \infty) \to \mathbb{R}$ be the function $f(x) = \sqrt{x}$. Construct the third Taylor approximant for $f$ about 4, and hence find an approximation to $f(5) = \sqrt{5}$. Use Taylor's Theorem to find upper and lower bounds on $\sqrt{5}$.

*Solution:* The first four derivatives of $f$ are

$$f'(x) = \frac{x^{-1/2}}{2},$$

$$f''(x) = -\frac{x^{-3/2}}{4},$$

$$f'''(x) = \frac{3x^{-5/2}}{8},$$

$$f^{(4)}(x) = -\frac{15x^{-7/2}}{16},$$

so $f(4) = 2$, $f'(4) = \frac{1}{4}$, $f''(4) = -2^{-5}$, and $f'''(4) = 3/2^8$. It follows that the third Taylor approximant for $f$ about 4 is

$$p_3(x) = f(4) + f'(4)(x - 4) + \frac{f''(4)}{2!}(x - 4)^2 + \frac{f'''(4)}{3!}(x - 4)^3$$

$$= 2 + \frac{1}{4}(x - 4) - \frac{1}{64}(x - 4)^2 + \frac{1}{512}(x - 4)^3.$$

Using the approximation $f(x) \approx p_3(x)$ gives

$$\sqrt{5} = f(5) \approx p_3(5) = 2 + \frac{1}{4} - \frac{1}{64} + \frac{1}{512} = 2.236328125.$$

By Taylor's Theorem, the true value of $f(5)$ is

$$f(5) = p_2(5) + \frac{f^{(4)}(c)}{4!}(5 - 4)^4 = p_2(5) - \frac{5}{2^7 c^{7/2}}$$

for some $c \in (4, 5)$. It follows that the true value of $\sqrt{5}$ is strictly less than $p_2(5)$, but strictly greater than $p_2(5) - 5/(2^7 4^{7/2})$. That is, we have established that

$$2.236022949 < \sqrt{5} < 2.236328125.$$

For comparison, my pocket calculator tells me that $\sqrt{5} \approx 2.236067977$. $\square$

## 9.5   Summary

- A function $f : D \to \mathbb{R}$ is **differentiable at** $a \in D$ if

$$\lim_{x \to a} \frac{f(x) - f(a)}{x - a}$$

  exists. In this case, we denote the limit $f'(a)$ and call it the **derivative** of $f$ at $a$. We say that $f$ is **differentiable** if it is differentiable at $a$ for all $a \in D$.
- The limit in $f'(a)$ is defined using sequences, as in Definition 8.1.
- Using this definition of $f'(a)$, we can prove that derivatives obey the usual rules of differential calculus, to wit:

  Linearity:       $(\alpha f + \beta g)'(a) = \alpha f'(a) + \beta g'(a)$,

  Product Rule:       $(fg)'(a) = f'(a)g(a) + f(a)g'(a)$,

  Quotient Rule:       $(f/g)'(a) = \dfrac{f'(a)g(a) - f(a)g'(a)}{g(a)^2}$,

  Chain Rule:       $(g \circ f)'(a) = g'(f(a))f'(a)$.

- The **Mean Value Theorem**: If $f : [a, b] \to \mathbb{R}$ is continuous, and differentiable on $(a, b)$, then there exists $c \in (a, b)$ such that

$$f'(c) = \frac{f(b) - f(a)}{b - a}.$$

- Let $f : I \to \mathbb{R}$ be differentiable, where $I \subseteq \mathbb{R}$ is an interval. We can use the Mean Value Theorem to prove that
  - If $f'(x) = 0$ for all $x \in I$ then $f$ is constant.
  - If $f'(x) > 0$ for all $x \in I$ then $f$ is strictly increasing.
  - If $f'(x) < 0$ for all $x \in I$ then $f$ is strictly decreasing.
- **Taylor's Theorem**: If $f$ is $(n + 1)$ times differentiable on some open interval $I$ and $a, x \in I$, then there exists $c$ between $a$ and $x$ such that

$$f(x) = f(a) + \frac{f'(a)}{1!}(x-a) + \cdots + \frac{f^{(n)}(a)}{n!}(x-a)^n + \frac{f^{(n+1)}(c)}{n!}(x-a)^{n+1}.$$

## 9.6   Tutorial problems

1 Compute the following derivatives directly from Definition 9.1:

   (a) $f'(3)$ where $f : \mathbb{R} \to \mathbb{R}$, $f(x) = x^2$.

   (b) $f'(1)$ where $f : (0, \infty) \to \mathbb{R}$, $f(x) = 1/x^2$.

   (c) $f'(0)$ where $f : \mathbb{R} \to \mathbb{R}$, $f(x) = \begin{cases} x^2 \sin(1/x) & \text{if } x \neq 0, \\ 0 & \text{if } x = 0. \end{cases}$

2 By careful use of the Chain Rule (Theorem 9.11) compute $f'(1)$ where $f : \mathbb{R} \to \mathbb{R}$, $f(x) = (1 + x^6)^{-7}$.

3 Let $f : A \to B$, $g : B \to C$, and $h : C \to \mathbb{R}$ be differentiable at $a \in A$, $f(a) \in B$, and $(g \circ f)(a) \in C$, respectively. Use the Chain Rule to show that $h \circ g \circ f : A \to \mathbb{R}$ is differentiable at $a$, and find a formula for $(h \circ g \circ f)'(a)$ in terms of the derivatives of $f$, $g$, and $h$.

4 Recall that a function $f : \mathbb{R} \to \mathbb{R}$ is **even** if $f(-x) = f(x)$ for all $x \in \mathbb{R}$, and is **odd** if $f(-x) = -f(x)$ for all $x \in \mathbb{R}$. Let $f : \mathbb{R} \to \mathbb{R}$ be differentiable. Show that

   (a) if $f$ is even then $f'$ is odd,

   (b) if $f$ is odd then $f'$ is even.

## 9.7   Homework problems

1 Compute the following derivatives directly from Definition 9.1:

   (a) $f'(-1)$ where $f : \mathbb{R} \to \mathbb{R}$, $f(x) = x^3$.

   (b) $f'(2)$ where $f : \mathbb{R} \to \mathbb{R}$, $f(x) = (x - 1)^2$.

   (c) $f'(0)$ where $f : \mathbb{R} \to \mathbb{R}$, $f(x) = \begin{cases} x^2 & \text{if } x \in \mathbb{Q}, \\ x^3 & \text{if } x \notin \mathbb{Q}. \end{cases}$

2 Find the maximum and minimum values attained by the function $f : [-1, 1] \to \mathbb{R}$, $f(x) = x^3 - x$. Rigorously justify your answers.

3 Prove from first principles that the function $f : [0, \infty) \to \mathbb{R}$, $f(x) = \sqrt{x}$, is differentiable on $(0, \infty)$, but not differentiable at 0.

4 Construct the third Taylor approximant of the function $f : \mathbb{R}\setminus\{-1\} \to \mathbb{R}$, $f(x) = 1/(x + 1)$ about 1. Use Taylor's Theorem to find a pair of degree four polynomial functions, $q(x)$ and $\bar{q}(x)$ such that

$$q(x) \leq f(x) \leq \bar{q}(x)$$

for all $x \in (0, 2)$.

# Chapter 10

# Power series

## 10.1  Definition and radius of convergence

A power series is a series of the form

$$\sum_{n=0}^{\infty} a_n x^n$$

where $x$ is interpreted as a real *variable*. The series depends on the choice of value for $x \in \mathbb{R}$. The terms of the series are $a_n x^n$, and the $k^{th}$ partial sum is

$$s_k = \sum_{n=0}^{k} a_n x^n = a_0 + a_1 x + a_2 x^2 + \cdots + a_k x^k,$$

just as for series in general. Note that, if $x = 0$, then every partial sum is $s_k = a_0$, so the series certainly converges (to $a_0$) in that case. In general, a power series may converge for some values of $x$ but diverge for others. If the series converges for a particular choice of $x$, its limit will, in general, depend on $x$. So a power series defines a real-valued function on (perhaps only part of) $\mathbb{R}$.

**Example 10.1.** Consider the series $\sum_{n=0}^{\infty} x^n$. Its $k^{\text{th}}$ partial sum is

$$s_k = 1 + x + x^2 + \cdots + x^k$$
$$\Rightarrow \quad x s_k = x + x^2 + x^3 + \cdots + x^{k+1}$$
$$\Rightarrow \quad (1 - x)s_k = 1 - x^{k+1}$$
$$\Rightarrow \quad s_k = \frac{1 - x^{k+1}}{1 - x}.$$

If $|x| < 1$ then $|x|^{k+1} \to 0$ (Example 3.15) so $s_k \to 1/(1 - x)$ (Algebra of Limits). Hence, for $|x| < 1$, the series converges to $1/(1-x)$. Note that this argument is just a verbatim repetition of Example 5.4. The only difference is that now we think of the "common ratio" as a real *variable*.

For $|x| \geq 1$, the terms of the series $x^n$ have $|x^n| \geq 1$, so certainly $x^n \not\to 0$, and we deduce that the series diverges (by the Divergence Test, Theorem 5.5).

In summary, if we use the power series to define a function

$$f(x) = \sum_{n=0}^{\infty} x^n$$

then $f : D \to \mathbb{R}$ where $D = (-1, 1)$ and for all $x \in D$,

$$f(x) = \frac{1}{1-x}.$$

$\square$

So we have another way of defining the function $f(x) = 1/(1-x)$, using power series. But what's the point of this? The power series definition has two obvious disadvantages

(i) It's more complicated. If we want to know what $f(1/2)$ is, it's much easier to compute

$$\frac{1}{1 - (1/2)} = 2$$

than to compute the infinite sum $\sum_{n=0}^{\infty} (1/2)^n$.

(ii) The power series only makes sense for $|x| < 1$, whereas $f(x) = 1/(1-x)$ is well-defined for all $x$ except 1.

Power series really come into their own in complex analysis (where we allow the variable $x$ to take values in the complex plane) and, in that setting, representing a function by a power series can be an extremely smart move indeed. But for our (real analysis) purposes, our main motivation for studying power series is that, for some functions, a definition in terms of power series is really the only definition we have.

**Definition 10.2.**    The **exponential, sine** and **cosine functions** are defined by

$$\exp : \mathbb{R} \to \mathbb{R}, \qquad \exp(x) = \sum_{n=0}^{\infty} \frac{x^n}{n!},$$

$$\sin : \mathbb{R} \to \mathbb{R}, \qquad \sin(x) = \sum_{n=0}^{\infty} (-1)^n \frac{x^{2n+1}}{(2n+1)!},$$

$$\cos : \mathbb{R} \to \mathbb{R}, \qquad \cos(x) = \sum_{n=0}^{\infty} (-1)^n \frac{x^{2n}}{(2n)!}.$$

Of course, we should check that these functions are well-defined, that is, that the power series defining them converge for all $x \in \mathbb{R}$ (unlike the power series for $f(x) = 1/(1-x)$). We'll check exp, and leave sin and cos as an exercise:

**Example 10.3.** Claim: $\exp(x) = \sum_{n=0}^{\infty} \frac{x^n}{n!}$ converges for all $x \in \mathbb{R}$.

**Proof.** As we've already noted, all power series converge at $x = 0$ (to $a_0 = 1/0! = 1$, in this case). So assume $x \neq 0$. Let $b_n = x^n/n!$. Then $|b_n| > 0$ and

$$\frac{|b_{n+1}|}{|b_n|} = \frac{|x|}{n+1} \to 0 < 1$$

so $\sum_{n=0}^{\infty} |b_n|$ converges, by the Ratio Test. Hence $\sum_{n=0}^{\infty} b_n$ converges (Theorem 5.20). $\square$

So, unlike the power series $\sum_{n=0}^{\infty} x^n$, which converges only for $x \in (-1, 1)$, the power series defining exp, sin, and cos converge for all $x \in \mathbb{R}$. On the other hand, it's not hard to come up with power series which converge only at $x = 0$.

**Example 10.4.** Claim: $\sum_{n=1}^{\infty} n^n x^n$ diverges for all $x \neq 0$.

**Proof.** Let $b_n = n^n x^n$. By the Archimedean Property of $\mathbb{R}$, there exists $N \in \mathbb{Z}^+$ such that $N \geq 2/|x|$. Then for all $n \geq N$

$$\frac{|b_{n+1}|}{|b_n|} = \frac{(n+1)^{n+1}}{n^n} \frac{|x|^{n+1}}{|x|^n} = |x|(n+1)\left(1 + \frac{1}{n}\right)^n$$
$$> |x|(n+1) > |x|n \geq 2.$$

Hence, for all $n \geq N$, $|b_n| \geq |b_N| 2^{n-N}$ (check by induction). So $|b_n|$ is unbounded above, and hence $b_n \not\to 0$. Hence, $\sum_{n=1}^{\infty} b_n$ diverges (by the Divergence Test). $\square$

Given a power series

$$\sum_{n=0}^{\infty} a_n x^n$$

it's interesting to ask "what is the subset $D \subseteq \mathbb{R}$ of values $x$ for which the series converges?" Since the coefficients $a_n$ could be absolutely any real numbers you like, it may seem that this set could be arbitrarily complicated. But actually, the examples we've already looked at illustrate (essentially) all possibilities: either

(i) the series converges for all $x$, or
(ii) it converges only for $x = 0$, or
(iii) there is some constant $R > 0$ such that the series converges for all $|x| < R$ and diverges for all $|x| > R$.

The only subtlety occurs in case (iii): the series may converge for $x = R$, $x = -R$, both, or neither.

**Definition 10.5.**   The **radius of convergence** of a power series $\sum_{n=0}^{\infty} a_n x^n$ is

$$R := \sup\left\{ |x| \ : \ x \in \mathbb{R}, \ \sum_{n=0}^{\infty} |a_n x^n| \text{ converges} \right\}.$$

Note the set $\{|x| : \sum_{n=0}^{\infty} |a_n x^n| \text{ converges}\} \subset \mathbb{R}$ always contains 0 (since $\sum_{n=0}^{\infty} |a_n x^n|$ converges to $|a_0|$ at $x = 0$), so is certainly nonempty. If the set is unbounded, we say $R = \infty$.

**Lemma 10.6.**   *Let $\sum_{n=0}^{\infty} a_n x^n$ converge at $x = x_1$. Then it converges absolutely for all $x$ such that $|x| < |x_1|$.*

**Proof.**   Since $\sum a_n x_1^n$ converges, $a_n x_1^n \to 0$ (by the Divergence Test), and so is bounded: there exists $K > 0$ such that for all $n$, $|a_n x_1^n| < K$. But then for all $n$,

$$|a_n x^n| = |a_n x_1^n| \frac{|x|^n}{|x_1|^n} < K \left( \frac{|x|}{|x_1|} \right)^n,$$

so if $|x| < |x_1|$, $\sum_{n=0}^{\infty} |a_n x^n|$ converges by comparison with the convergent (geometric) series $\sum_{n=0}^{\infty} (|x|/|x_1|)^n$. $\qquad\square$

**Theorem 10.7.**   *Let $\sum_{n=0}^{\infty} a_n x^n$ have radius of convergence $R$. Then*

$$\sum_{n=0}^{\infty} a_n x^n \begin{cases} \text{converges absolutely for } |x| < R, \\ \text{diverges for } |x| > R. \end{cases}$$

**Proof.**   Let $A := \{|x| : \sum_{n=0}^{\infty} |a_n x^n| \text{ converges}\} \subset \mathbb{R}$, so that $R = \sup A$.

If $|x| < R$ then there exists $x_1$ with $|x| < |x_1| < R$ such that $\sum_{n=0}^{\infty} a_n x_1^n$ is absolutely convergent (else $R$ isn't the *least* upper bound on $A$, $|x|$ being smaller), hence convergent, so $\sum_{n=0}^{\infty} a_n x^n$ is absolutely convergent by Lemma 10.6.

If $|x| > R$ and $\sum_{n=0}^{\infty} a_n x^n$ converges then $\sum_{n=0}^{\infty} a_n x_2^n$, where $x_2 = (|x| + R)/2$, converges absolutely by Lemma 10.6, since $|x_2| < |x|$. But $|x_2| > R$, a contradiction ($R$ isn't an upper bound on $A$ at all!), so we conclude that $\sum_{n=0}^{\infty} a_n x^n$ diverges. $\qquad\square$

So, given a power series, if we can figure out its radius of convergence $R$, we know immediately that it converges absolutely on the open interval $(-R, R)$ and diverges for all $|x| > R$. This is very useful information, and it's important to be able to compute radii of convergence. Luckily, this is usually possible by a simple application of the Ratio Test.

**Example 10.8.**

(i) What is the radius of convergence of $\sum_{n=0}^{\infty} \dfrac{n^2}{n+1} x^{2n}$?

*Solution:* Let $b_n = \dfrac{n^2}{n+1} x^{2n}$. Then, for all $x \neq 0$ and $n > 0$, $|b_n| > 0$, so we can (try to) use the Ratio Test to see whether $\sum |b_n|$ converges. Now

$$\frac{|b_{n+1}|}{|b_n|} = \frac{(n+1)^2}{n+2} |x|^{2n+2} \frac{n+1}{n^2 |x|^{2n}}$$

$$= \frac{(n+1)^3}{n^2(n+2)} |x|^2$$

$$= \frac{(1 + (1/n))^3}{1 + (2/n)} |x|^2$$

$$\to |x|^2.$$

So if $|x| < 1$, $|b_{n+1}|/|b_n|$ converges to a limit less than 1, so $\sum |b_n|$ converges (by the Ratio Test), that is $\sum b_n$ converges absolutely. On the other hand, for $|x| > 1$, $\sum b_n$ does *not* converge absolutely (again, by the Ratio Test). Comparing with Definition 10.5, we deduce that $R = 1$.

(ii) What is the radius of convergence of $\sum_{n=1}^{\infty} \dfrac{n^3}{8^n} x^{3n-1}$?

*Solution:* Let $b_n = \dfrac{n^3}{8^n} x^{3n-1}$. Then, for all $x \neq 0$, $|b_n| > 0$, so we can (try to) use the Ratio Test to see whether $\sum |b_n|$ converges. Now

$$\frac{|b_{n+1}|}{|b_n|} = \frac{(n+1)^3}{n^3} \frac{8^n}{8^{n+1}} |x|^3$$

$$= \left(1 + \frac{1}{n}\right)^3 \frac{1}{8} |x|^3$$

$$\to \frac{|x|^3}{8}.$$

So by the Ratio Test, $\sum b_n$ converges absolutely if $|x| < 2$, but not if $|x| > 2$. Comparing with Definition 10.5, we deduce that $R = 2$. $\square$

**Warning!** Example 10.8 describes a method of *finding* the radius of convergence, not the *definition* of the radius of convergence. Some people may try and tell you that the radius of convergence of a power series $\sum_{n=0}^{\infty} a_n x^n$ is $R = \lim_{n \to \infty} |a_n|/|a_{n+1}|$, a formula which results from (blindly) applying the method of Example 10.8 to a "general" power series, without thinking carefully about whether it works. Have no truck with such charlatans! Note that, for both the power series we just considered, the formula generates gibberish:

$$\sum_{n=0}^{\infty} \frac{n^2}{n+1} x^{2n} \quad \text{has} \quad a_n = \begin{cases} 0, & n \text{ odd}, \\ \frac{n^2}{n+1}, & n \text{ even}, \end{cases}$$

so for all even $n$, $|a_n|/|a_{n+1}|$ is undefined, and it is meaningless to speak of the limit of this sequence. Similarly

$$\sum_{n=0}^{\infty} \frac{n^3}{8^n} x^{3n-1} \quad \text{has} \quad a_n = \begin{cases} 0, & n+1 \text{ not divisible by 3}, \\ \frac{n^3}{8^n}, & n+1 \text{ divisible by 3}, \end{cases}$$

so again the sequence $|a_n|/|a_{n+1}|$ is undefined. But of course, both these power series had perfectly well-defined radii of convergence, and the method of Example 10.8, applied carefully, allowed us to compute them. The moral of the story is this: don't confuse the *definition* of a mathematical object with a *method* used to find that mathematical object. The definition of radius of convergence is Definition 10.5 and it makes no mention of the sequence $|a_n|/|a_{n+1}|$, or the Ratio Test.

So we have now defined the functions exp, sin, and cos, and we know that our definitions make sense on the whole real line. Our next order of business is to show that these functions have the properties we know and love from our previous (informal) study of them: that they are smooth, that exp is strictly increasing, that the derivative of sin is cos, etc. We begin by considering the *differentiability* of power series in general.

## 10.2   Differentiability of power series

Given a power series

$$f(x) = \sum_{n=0}^{\infty} a_n x^n$$

its partial sums can be considered to be a sequence of polynomial functions:

$$s_0(x) = a_0$$
$$s_1(x) = a_0 + a_1 x$$
$$s_2(x) = a_0 + a_1 x + a_2 x^2$$
$$\vdots$$

Each of these is certainly a differentiable function (on the whole of $\mathbb{R}$), so one might expect that their limit, $f(x) = \lim_{k \to \infty} s_k(x)$, is also a differentiable function, and that

$$f'(x) = \lim_{k \to \infty} s_k'(x) = \lim_{k \to \infty} a_1 + x_2 x + \cdots k a_k x^{k-1} = \sum_{n=1}^{\infty} n a_n x^{n-1}.$$

This turns out to be true, at least on $(-R, R)$ where $R$ is the radius of convergence of $f$, but proving it is no simple matter. The problem is that $f'(x)$ is really a *double* limit

$$f'(x) = \lim_{y \to x} \frac{f(x) - f(y)}{x - y} = \lim_{y \to x} \lim_{k \to \infty} \sum_{n=0}^{k} a_n \frac{y^n - x^n}{y - x},$$

and our expected formula for $f'(x)$ results from computing the double limit in the *opposite order*

$$\lim_{k \to \infty} \lim_{y \to x} \sum_{n=0}^{k} a_n \frac{y^n - x^n}{y - x}.$$

But, as we have seen previously (Example 8.9), swapping the order in which we evaluate a double limit can, in general, change its value, so computing the second limit tells us precisely nothing about the first. We will need to work considerably harder!

We begin by showing that a power series and the obvious candidate for its derivative have the same radius of convergence.

**Lemma 10.9.** *The power series* $f(x) = \sum_{n=0}^{\infty} a_n x^n$ *and* $g(x) = \sum_{n=1}^{\infty} n a_n x^{n-1}$ *have the same radius of convergence.*

**Proof.** We define the following subsets of $\mathbb{R}$,

$$A = \{|x| : \sum |a_n x^n| \text{ converges}\}, \qquad B = \{|x| : \sum |n a_n x^{n-1}| \text{ converges}\},$$

and recall that $\sup A$ and $\sup B$ are the radii of convergence of $f$ and $g$, respectively. We will prove that

(i) for all $x$ with $|x| < \sup B$, $f(x)$ converges absolutely (so $|x| \in A$, and hence $\sup A \geq \sup B$), and

(ii) for all $x$ with $|x| < \sup A$, $g(x)$ converges absolutely (so $|x| \in B$, and hence $\sup B \geq \sup A$).

It follows immediately that $\sup A = \sup B$, which is what we seek to prove.

(i) Let $x \in \mathbb{R}$ with $|x| < \sup B$. Then $s_k = \sum_{n=1}^{k} |na_n x^{n-1}|$ converges (Theorem 10.7), and hence is bounded above, and

$$t_k = \sum_{n=0}^{k} |a_n x^n| = |a_0| + |x| \sum_{n=1}^{k} |a_n x^{n-1}| \leq |a_0| + |x| s_k.$$

So $(t_k)$ is increasing and bounded above, and hence converges, by the Monotone Convergence Theorem. Hence, $|x| \in A$.

(ii) Let $x \in \mathbb{R}$ with $|x| < \sup A$. If $x = 0$ then $g(x)$ certainly converges, so we may assume $|x| > 0$. Choose $\rho \in (|x|, \sup A)$. Then $t_k = \sum_{n=0}^{k} |a_n| \rho^n$ converges (Theorem 10.7), and hence is bounded above. Now

$$s_k = \sum_{n=1}^{k} |na_n x^{n-1}| = |x|^{-1} \sum_{n=1}^{\infty} n \left( \frac{|x|}{\rho} \right)^n |a_n| \rho^n.$$

We have seen that the sequence $n(|x|/\rho)^n \to 0$, since $(|x|/\rho) < 1$ (Example 5.13), so it must be bounded, that is, there exists $K > 0$ such that, for all $n$, $n(|x|/\rho)^n \leq K$. But then

$$s_k \leq |x|^{-1} \sum_{n=0}^{k} K|a_n| \rho^n = \frac{K}{|x|} t_k.$$

Hence, $s_k$ is increasing and bounded above, so converges, by the Monotone Convergence Theorem (Corollary 3.14). Hence, $|x| \in B$. $\qquad \square$

We now show that $g(x)$ really is the derivative of $f(x)$.

**Theorem 10.10.** *Suppose that $f(x) = \sum_{n=0}^{\infty} a_n x^n$ has radius of convergence $R > 0$. Then $f$ is differentiable on $(-R, R)$, and*

$$f'(x) = \sum_{n=1}^{\infty} na_n x^{n-1}.$$

**Proof.** Choose and fix $y \in (-R, R)$. Let

$$g(x) = \sum_{n=1}^{\infty} na_n x^{n-1}, \quad \text{and} \quad h(x) = \sum_{n=2}^{\infty} n(n-1)a_n x^{n-2}.$$

By Lemma 10.9, $g(x)$ has radius of convergence $R$, so by Lemma 10.9 again (applied to $g(x)$), $h(x)$ also has radius of convergence $R$. Let $(x_j)$ be any sequence in $(-R, R)\backslash\{y\}$ converging to $y$. Then the sequence

$$s_j = \frac{f(x_j) - f(y)}{x_j - y} - g(y)$$

is well-defined, and we seek to show that $s_j \to 0$.

By Theorem 10.7, the power series $h(x)$ converges absolutely on $(-R, R)$. Hence, the function

$$H : (0, R) \to \mathbb{R}, \qquad H(r) = \sum_{n=2}^{\infty} n(n-1)|a_n|r^{n-2}$$

is well-defined and is manifestly increasing. We will show that there exists $r_* \in (0, R)$ such that $|s_j| \leq \frac{1}{2}H(r_*)|x_j - y|$ for all $j$ sufficiently large. It then follows that $s_j \to 0$ (by the Squeeze Rule).

For each $j, k \in \mathbb{Z}^+$, let

$$P_{jk} = \frac{1}{x_j - y} \sum_{n=1}^{k} a_n(x_j^n - y^n) - \sum_{n=1}^{k} n a_n y^{n-1}$$

and note that $s_j = \lim_{k \to \infty} P_{jk}$. Repeatedly using the identity

$$x^p - y^p = (x - y)(x^{p-1} + x^{p-2}y + \cdots + xy^{p-2} + y^{p-1})$$

we see that

$$P_{jk} = \sum_{n=1}^{k} a_n\{x_j^{n-1} + x_j^{n-2}y + \cdots + x_j y^{n-2} + y^{n-1} - n y^{n-1}\}$$

$$= \sum_{n=2}^{k} a_n\{(x_j^{n-1} - y^{n-1}) + (x_j^{n-2}y - y^{n-1}) + \cdots$$

$$\cdots + (x_j y^{n-2} - y^{n-1}) + (y^{n-1} - y^{n-1})\}$$

$$= (x_j - y) \sum_{n=2}^{k} a_n\{(x_j^{n-2} + x_j^{n-3}y + \cdots + x_j y^{n-3} + y^{n-2})$$

$$+ y(x_j^{n-3} + x_j^{n-4}y + \cdots + x_j y^{n-4} + y^{n-3})$$

$$+ y^2(x_j^{n-4} + x_j^{n-5}y + \cdots + x_j y^{n-5} + y^{n-4}) + \cdots$$

$$\cdots + y^{n-2}\}$$

$$= (x_j - y) \sum_{n=2}^{k} a_n\{x_j^{n-2} + 2y x_j^{n-3} + 3y^2 x_j^{n-4} + \cdots + (n-1)y^{n-2}\}.$$

Let $r_j = \max\{|x_j|, |y|\} \in (0, R)$. Then

$$|P_{jk}| \leq |x_j - y| \sum_{n=2}^{k} |a_n|(1 + 2 + \cdots + (n-1))r_j^{n-2}$$

$$= \frac{1}{2}|x_j - y| \sum_{n=2}^{k} n(n-1)|a_n|r_j^{n-2}$$

$$\leq \frac{1}{2}|x_j - y|H(r_j).$$

Now $x_j \to y$, so $r_j \to |y| < R$, and hence there exists $J \in \mathbb{Z}^+$ such that, for all $j \geq J$, $r_j < r_* = \frac{1}{2}(|y| + R)$. Then, since $H$ is an increasing function, for all $j \geq J$, $H(r_j) \leq H(r_*)$, and hence,

$$|P_{jk}| \leq \frac{1}{2}H(r_*)|x_j - y|.$$

Now, for each fixed $j \geq J$, the sequence $|P_{jk}|$ indexed by $k$ certainly converges, to $|s_j|$, and each term in this sequence is no greater than the constant (independent of $k$) $\frac{1}{2}H(r_*)|x_j - y|$ so, by Proposition 3.4,

$$|s_j| \leq \frac{1}{2}H(r_*)|x_j - y|.$$

This inequality holds for all $j \geq J$ and so, by the Squeeze Rule, $s_{j+J} \to 0$, so $s_j \to 0$ (by the Tail Lemma) as was to be shown. $\qquad\square$

So the derivative of a power series exists on its open interval of convergence and is just the power series obtained by termwise differentiation. Since $f'(x)$ is also a power series with radius of convergence $R$, we can apply Theorem 10.10 to $f'(x)$ and deduce that $f : (-R, R) \to \mathbb{R}$ is actually *twice* differentiable. Further, $f''(x)$ also has radius of convergence $R$, so is differentiable on $(-R, R)$, that is, $f$ is *three times* differentiable. In fact, we can keep applying Theorem 10.10 as often as we like, and we conclude that $f$ is *smooth*.

**Corollary 10.11.** *Let $f(x) = \sum_{n=0}^{\infty} a_n x^n$ have radius convergence $R > 0$. Then $f : (-R, R) \to \mathbb{R}$ is a smooth function, and*

$$a_n = \frac{f^{(n)}(0)}{n!}.$$

**Proof.** For each integer $k \geq 0$, define

$$g_k(x) = \sum_{n=0}^{\infty} \frac{(n+k)!}{n!}a_{n+k}x^n.$$

I claim that each power series $g_k(x)$ has radius of convergence $R$ and that, for all $x \in (-R, R)$, $f^{(k)}(x) = g_k(x)$. We prove this by induction on $k$.

Certainly the claim holds for $k = 0$, since $g_0 = f$. So, assume that the claim holds for some value $k \geq 0$, and consider $g_{k+1}$. Defining coefficients $b_n = \frac{(n+k)!}{n!} a_{n+k}$ so that $g_k(x) = \sum_{n=0}^{\infty} b_n x^n$, then

$$g_{k+1}(x) = \sum_{n=0}^{\infty} \frac{(n+k+1)!}{n!} a_{n+k+1} x^n = \sum_{m=1}^{\infty} \frac{(m+k)!}{(m-1)!} a_{m+k} x^{m-1}$$

$$= \sum_{m=1}^{\infty} m b_m x^{m-1}.$$

Hence, by Lemma 10.9, $g_{k+1}$ has the same radius of convergence as $g_k$, and by Theorem 10.10, $g_{k+1}(x)$ coincides with the derivative of $g_k$. But, by our induction hypothesis, $g_k$ has radius of convergence $R$ and coincides with $f^{(k)}$, so $g_{k+1}$ has radius of convergence $R$ and coincides with $f^{(k+1)}$. Hence, if the claim holds for some $k \geq 0$, it also holds for $k + 1$. Hence, by induction, the claim holds for all integers $k \geq 0$.

It follows that $f^{(k)}(0) = g_k(0) = k! a_k$, which completes the proof. □

Corollary 10.11 applies to any power series with non-zero radius of convergence so, in particular, it applies to the series defining exp, sin, and cos.

**Proposition 10.12.** *The functions* $\exp : \mathbb{R} \to \mathbb{R}$, $\sin : \mathbb{R} \to \mathbb{R}$, $\cos : \mathbb{R} \to \mathbb{R}$ *defined in Definition 10.2 are smooth, and their derivatives are*

$$\exp' = \exp, \qquad \sin' = \cos, \qquad \cos' = -\sin.$$

**Proof.** That these functions are smooth follows immediately from Corollary 10.11. Furthermore, by Theorem 10.10

$$\exp'(x) = \sum_{n=1}^{\infty} n \left( \frac{1}{n!} \right) x^{n-1} = \sum_{n=1}^{\infty} \frac{x^{n-1}}{(n-1)!} = \sum_{m=0}^{\infty} \frac{x^m}{m!} = \exp(x)$$

for all $x \in \mathbb{R}$, so $\exp' = \exp$. The formulae for $\sin'$ and $\cos'$ follow from similar arguments. □

## 10.3 Properties of the exponential function

In this section we will prove that the function $\exp : \mathbb{R} \to \mathbb{R}$, defined as a power series in Definition 10.2, has all the properties you're familiar with. We begin with the following fundamental identity.

**Lemma 10.13.** *For all $x, y \in \mathbb{R}$, $\exp(x + y) = \exp(x) \exp(y)$.*

**Proof.** For each fixed number $b \in \mathbb{R}$, define the function $f : \mathbb{R} \to \mathbb{R}$,

$$f(x) = \exp(x) \exp(b - x).$$

This is a product of differentiable functions, so is differentiable, and by the Product and Chain Rules, and Proposition 10.12,

$$f'(x) = \exp'(x) \exp(b - x) - \exp(x) \exp'(b - x) = 0.$$

Since $f'(x) = 0$ for all $x$ in the interval $\mathbb{R}$, it follows that $f$ is constant (Corollary 9.21). Hence, for all $x \in \mathbb{R}$ and any $b \in \mathbb{R}$, $f(x) = f(b)$, that is,

$$\exp(x) \exp(b - x) = \exp(b).$$

Applying this in the case where $b = x + y$ establishes the claim. $\square$

This lemma explains why $\exp(x)$ is often denoted $e^x$ and thought of as the constant $e = \exp(1) = \sum_{n=0}^{\infty} \frac{1}{n!}$ "raised to the power $x$". The point is that, written in this alternative notation, Lemma 10.13 looks like one of the standard algebraic rules of integer exponents:

$$e^{x+y} = e^x \times e^y \qquad \text{in analogy with} \qquad a^{n+m} = a^n \times a^m.$$

This analogy, and the associated notation, are a useful mnemonic, but it's important to realize that they are just that, an analogy. We don't literally *define* $\exp(x)$ to be the irrational number $e$ "raised to the power $x$", for obvious reasons: what on earth does it mean to raise $e$ to the power $\sqrt{2}$, for example? In fact, the logic works in the opposite direction. Having proved that the function exp, defined as a power series, satisfies a rule analogous to the behaviour of integer (and rational) exponents, we *define* $e^x$, for any exponent $x$ (rational or irrational) to be $\exp(x)$. We will see later (section 12.2) how this allows us to define $a^x$ for any $a > 0$ and $x \in \mathbb{R}$.

**Proposition 10.14.** *The function* exp *is a bijective mapping* $\mathbb{R} \to (0, \infty)$.

**Proof.** We first note that, for all $x \in \mathbb{R}$, $\exp(x) \neq 0$ since, if there exists $x \in \mathbb{R}$ such that $\exp(x) = 0$, then Lemma 10.13 in the case $y = -x$ implies that

$$1 = \exp(0) = 0 \times \exp(-x) = 0.$$

Now $\exp(0) = 1 > 0$, and exp is differentiable, hence continuous, so it follows from the Intermediate Value Theorem that $\exp(x) > 0$ for all $x \in \mathbb{R}$ (if $\exp(x) < 0$ then there exists $c$ between $x$ and $0$ such that $\exp(c) = 0$).

Hence exp maps $\mathbb{R}$ to $(0, \infty)$. Furthermore, for all $x$, $\exp'(x) = \exp(x) > 0$, so exp is strictly increasing (Proposition 9.24), and hence is injective.

It remains to show that $\exp : \mathbb{R} \to (0, \infty)$ is surjective. Note that if $\exp(x) = y$ then, by Lemma 10.13 $\exp(-x)y = \exp(0) = 1$, that is, $\exp(-x) = 1/y$, so it suffices to show that exp takes all values in $[1, \infty)$. Let any $y \in [1, \infty)$ be given. Then

$$\exp(y) = \sum_{n=0}^{\infty} \frac{y^n}{n!} > 1 + \frac{y}{1!} > y$$

and $\exp(0) = 1 \leq y$. So $y$ is a number between $\exp(0)$ and $\exp(y)$, and exp is continuous, so there exists some $x$ between $0$ and $y$ such that $\exp(x) = y$. $\qquad\square$

Since $\exp : \mathbb{R} \to (0, \infty)$ is bijective, it has a well-defined inverse function $(0, \infty) \to \mathbb{R}$. This function is called the **natural logarithm**. We will study it further, from a very different viewpoint, in Chapter 12.

## 10.4 Elementary properties of the trigonometric functions

We have defined the trigonometric functions sine and cosine using power series

$$\sin x = \sum_{n=0}^{\infty} (-1)^n \frac{x^{2n+1}}{(2n+1)!}, \qquad \cos x = \sum_{n=0}^{\infty} (-1)^n \frac{x^{2n}}{(2n)!}.$$

This is in contrast to the way in which sin and cos are usually (informally) introduced to students of calculus, using the geometry of right angled triangles or of the unit circle. The power series definition has the advantage of being mathematically precise and unambiguous, unlike the geometric "definition" (what, precisely, is an "angle", for example?). It also has the advantage that we can apply all the technical tools we developed for power series in general to sin and cos. This allowed us to deduce immediately that both functions are smooth and that $\sin' = \cos$ and $\cos' = -\sin$, for example. However, many other features of sin and cos which are immediately apparent from their geometric "definition" are completely hidden when one writes them as power series. It is not clear, for example, that sin and cos are both periodic, with equal period, or that, for all $x \in \mathbb{R}$, $\sin^2 x + \cos^2 x = 1$. The purpose of this section is to prove, rigorously, that the functions sin and cos, defined by the power series above, have all the properties that you're familiar with. This retrospectively justifies all the assumptions we have made about these functions in earlier chapters, where we used them to generate interesting examples of sequences, functions, and limits. We begin with the simplest trigonometric identity of them all.

**Lemma 10.15.** *For all* $x \in \mathbb{R}$, $\sin^2 x + \cos^2 x = 1$.

**Proof.** Let $f : \mathbb{R} \to \mathbb{R}$ be the function $f(x) = \sin^2 x + \cos^2 x$. Then $f$ is differentiable, and by the Product Rule and Proposition 10.12,

$$f'(x) = 2\sin x \cos x + 2\cos x(-\sin x) = 0.$$

Hence, by Corollary 9.21, $f$ is contant, that is, for all $x \in \mathbb{R}$, $f(x) = f(0) = 0^2 + 1^2 = 1$. □

By a similar argument, we can prove two other useful identities.

**Lemma 10.16.** *For all* $x, y \in \mathbb{R}$,

(i) $\cos(x + y) = \cos x \cos y - \sin x \sin y$,
(ii) $\sin(x + y) = \sin x \cos y + \cos x \sin y$.

**Proof.** Let $y \in \mathbb{R}$ be a fixed constant, and define the function $f : \mathbb{R} \to \mathbb{R}$ by

$$f(x) = \cos(x + y) - \cos x \cos y + \sin x \sin y.$$

Now, by the Product Rule and Proposition 10.12,

$$f'(x) = -\sin(x + y) + \sin x \cos y + \cos x \sin y$$
$$f''(x) = -\cos(x + y) + \cos x \cos y - \sin x \sin y = -f(x).$$

We seek to show that, for all $x \in \mathbb{R}$, (i) $f(x) = 0$ and (ii) $f'(x) = 0$. Note that $f(0) = \cos y - \cos y + 0 = 0$ and $f'(0) = -\sin y + 0 + \sin y = 0$. Consider the function $g : \mathbb{R} \to \mathbb{R}$,

$$g(x) = \frac{1}{2}(f(x)^2 + f'(x)^2).$$

This is differentiable, and

$$g'(x) = f(x)f'(x) + f'(x)f''(x) = f(x)f'(x) + f'(x)(-f(x)) = 0$$

for all $x \in \mathbb{R}$, so by Corollary 9.21, $g$ is constant. Hence, for all $x \in \mathbb{R}$,

$$g(x) = g(0) = \frac{1}{2}(0^2 + 0^2) = 0.$$

It follows that, $f(x)^2 + f'(x)^2 = 0$ for all $x$, whence $f(x) = f'(x) = 0$. □

Proving that sin and cos are periodic with equal period is not so straightforward. We begin by showing that sin takes at least one negative value.

**Lemma 10.17.** $\sin 4 < 0$.

**Proof.** By definition, $\sin 4 = \lim_{k \to \infty} s_k$ where

$$s_k = \sum_{n=0}^{k} (-1)^n \frac{4^{2n+1}}{(2n+1)!}.$$

Since $s_k \to \sin 4$, the subsequence $s_{2m} \to \sin 4$ also. Now, for all $m \geq 2$,

$$s_{2m} = s_3 + \sum_{n=4}^{2m} (-1)^n \frac{4^{2n+1}}{(2n+1)!}$$

$$< s_3 + \sum_{p=2}^{m} \frac{4^{4p+1}}{(4p+1)!}$$

where we've obtained this estimate by omitting all the negative terms, with $n$ odd, from the sum. I claim that, for all $p \geq 2$,

$$\frac{4^{4p+1}}{(4p+1)!} < \frac{3}{2^{4p-6}}.$$

(Check this by induction: it's true for $p = 2$, and if it's true for $p = q \geq 2$, then

$$\frac{4^{4(q+1)+1}}{(4(q+1)+1)!} = \frac{64}{(4q+5)(4q+4)(4q+3)(4q+2)} \frac{4^{4q+1}}{(4q+1)!}$$

$$< \frac{64}{(4q)^4} \frac{3}{2^{4q-6}}$$

$$\leq \frac{1}{64} \frac{3}{2^{4q-6}}$$

$$< \frac{3}{2^{4(q+1)-6}},$$

so it's also true for $p = q + 1$.) Hence, for all $m \geq 2$,

$$s_{2m} < s_3 + \sum_{p=2}^{m} \frac{3}{2^{4p-6}} = s_3 + \frac{3}{4} \sum_{q=0}^{m-2} \left(\frac{1}{2^4}\right)^q =: t_m.$$

But, as we saw in Example 5.4,

$$t_m \to s_3 + \frac{3}{4} \frac{1}{1 - 1/2^4} = -\frac{184}{315}.$$

Hence,

$$\sin 4 = \lim_{m \to \infty} s_{2m} \leq -\frac{184}{315} < 0.$$

$\square$

**Lemma 10.18.** *There exists $x \in (0,1)$ such that $\sin x > 0$.*

**Proof.** Asume, to the contrary, that for all $x \in (0,1)$, $\sin x \leq 0$. Let $(x_n)$ be any sequence in $(0,1)$ such that $x_n \to 0$. Then

$$y_n = \frac{\sin x_n}{x_n} = \frac{\sin x_n - \sin 0}{x_n - 0} \to \sin'(0) = \cos 0 = 1.$$

But, by assumption, $y_n \leq 0$ for all $n$, contradicting Proposition 3.4. $\quad\square$

**Lemma 10.19.** *There exists $x \in (0,4)$ such that $\sin x = 1$.*

**Proof.** By the Extreme Value Theorem, $\sin : [0,4] \to \mathbb{R}$ attains a maximum value. By Lemma 10.18, there exists $p \in (0,4)$ with $\sin p > 0$, so this maximum value is not attained at 0 (since $\sin 0 = 0$), nor is it attained at 4 (since $\sin 4 < 0$, Lemma 10.17). Hence, the maximum value is attained at some interior point $x \in (0,4)$. By the Interior Extremum Theorem,

$$\sin'(x) = \cos x = 0,$$

so, by Lemma 10.15, $\sin^2 x = 1$. Clearly $\sin x \neq -1$ (as the maximum value is positive), so $\sin x = 1$. $\quad\square$

**Lemma 10.20.** *Let $A = \{x \in (0,\infty) : \sin x = 1\}$. Then $A$ has a minimum element.*

**Proof.** By Lemma 10.19, $A$ is nonempty, and $A$ is clearly bounded below, so, by Proposition 1.23, $A$ has an infimum, call it $K$ say. Clearly $K \geq 0$. Now, for each $n \in \mathbb{Z}^+$, $K + \frac{1}{n} > K$, so is not a lower bound on $A$. Hence, there exists $x_n \in A$ such that $K \leq x_n < K + \frac{1}{n}$. By the Squeeze Rule, $x_n \to K$. Now sin is continuous, so $\sin x_n \to \sin K$. But $x_n \in A$, so by definition, $\sin x_n = 1$ for all $n$. Hence $\sin K = 1$, so $K > 0$, and $K \in A$. Since $K$ is the infimum of $A$, and lies in $A$, it is the minimum element of $A$. $\quad\square$

We can use Lemma 10.20 to *define* the number $\pi$: it's the smallest positive real number such that $\sin(\pi/2) = 1$.

**Definition 10.21.** We define the real constant $\pi$ to be $2 \min A$. This exists, by Lemma 10.20.

**Theorem 10.22.** *For all $x \in \mathbb{R}$,*

*(i)* $\sin(x + \frac{\pi}{2}) = \cos x$
*(ii)* $\cos(x + \frac{\pi}{2}) = -\sin x$.

**Proof.** (i) By Lemma 10.16,

$$\sin(x + \frac{\pi}{2}) = \sin x \cos \frac{\pi}{2} + \cos x \sin \frac{\pi}{2}.$$

By the definition of $\pi$, $\sin \frac{\pi}{2} = 1$ so, by Lemma 10.15, $\cos \frac{\pi}{2} = 0$, and the result immediately follows.

(ii) Similarly, by Lemma 10.16,

$$\cos(x + \frac{\pi}{2}) = \cos x \times 0 - \sin x \times 1 = -\sin x.$$

$\square$

It follows immediately from Theorem 10.22 that sin and cos are *antiperiodic* with period $\pi$, that is,

$$\sin(x + \pi) = \cos(x + \frac{\pi}{2}) = -\sin x$$
$$\cos(x + \pi) = -\sin(x + \frac{\pi}{2}) = -\cos x$$

and hence are periodic with period $2\pi$,

$$\sin(x + 2\pi) = \sin x, \qquad \cos(x + 2\pi) = \cos x.$$

## 10.5   Summary

- A **power series** is a series of the form

$$\sum_{n=0}^{\infty} a_n x^n$$

where $x$ is a real variable. Its **radius of convergence** is

$$R = \sup\{|x| \ : \ \textstyle\sum_{n=0}^{\infty} a_n x^n \text{ converges absolutely}\}.$$

- If a power series has radius of convergence $R$, it converges absolutely for all $|x| < R$ and diverges for all $|x| > R$.
- A power series $f(x) = \sum_{n=0}^{\infty} a_n x^n$ with radius of convergence $R > 0$ is **differentiable** on $(-R, R)$ and its derivative is

$$f'(x) = \sum_{n=1}^{\infty} n a_n x^{n-1}.$$

This power series also has radius of convergence $R$.

- Convergent power series are, in fact, **smooth**.
- The exponential, sine, and cosine functions are defined by the power series

$$\exp(x) = \sum_{n=0}^{\infty} \frac{x^n}{n!},$$

$$\sin x = \sum_{n=0}^{\infty} (-1)^n \frac{x^{2n+1}}{(2n+1)!},$$

$$\cos x = \sum_{n=0}^{\infty} (-1)^n \frac{x^{2n}}{(2n)!}.$$

These converge for all $x \in \mathbb{R}$ and have the properties one expects.

## 10.6 Tutorial problems

1. Compute the radius of convergence of the following power series

$$\text{(a)} \sum_{n=0}^{\infty} \frac{nx^{4n}}{2^{2n+1}}, \quad \text{(b)} \sum_{n=0}^{\infty} \frac{n!}{2^{n(n+1)}} x^{(n^2)}.$$

2. Show that, for all $x \in (-1, 1)$,

$$\sum_{n=1}^{\infty} nx^{n-1} = \frac{1}{(1-x)^2}.$$

Hence compute the limit of the convergent series $\sum_{n=1}^{\infty} \frac{n+1}{2^n}$.

## 10.7 Homework problems

1. Compute the radius of convergence of the following power series.

$$\text{(a)} \sum_{n=0}^{\infty} 2^n x^{2n}, \quad \text{(b)} \sum_{n=2}^{\infty} \frac{(1+n)}{(1-n)^n} x^{2n+1}, \quad \text{(c)} \sum_{n=0}^{\infty} n! x^n.$$

2. Consider the power series $f(x) = \sum_{n=1}^{\infty} (-1)^{n+1} \frac{x^n}{n}$.

   (a) Show that its radius of convergence is $R = 1$.
   (b) It follows that $f$ defines a smooth function $f : (-1, 1) \to \mathbb{R}$. Show that $f'(x) = 1/(1+x)$.
   (c) Consider the function $g : (-\infty, 0] \to \mathbb{R}$, $g(x) = f(e^x - 1)$. Show that $g(x) = x$.

# Chapter 11

# Integration

## 11.1 Dissections and Riemann sums

How do we define the area of a bounded subset of the plane $\mathbb{R}^2$? If the subset is a rectangle, the answer is easy: its area is its length times its width. Similarly, if the subset is a union of non-overlapping rectangles it's easy: we just add up the areas of all the constituent rectangles. But what if the subset is more complicated: the region bounded by the $x$-axis, the vertical lines $x = a$ and $x = b > a$ and the graph $y = f(x)$ of some non-constant function, for example? One approach is to define the area of such a region to be the unique real number (if it exists) which is no bigger than the total area of any collection of rectangles which covers the region, and no smaller than the total area of any collection of rectangles which is covered by the region. This is the underlying idea that leads to the *Riemann integral*. We begin by identifying the collections of rectangles we will use. These are determined by *dissecting* the interval $[a, b]$ into a finite collection of subintervals.

**Definition 11.1.** A **dissection** of a closed bounded interval $[a, b]$ is a finite subset $\mathscr{D}$ of $[a, b]$ containing both $a$ and $b$. By convention, if $\mathscr{D}$ has $n + 1$ elements, we label these $a_0, a_1, \ldots, a_n$, so that

$$a = a_0 < a_1 < a_2 < \cdots < a_n = b,$$

and say that $\mathscr{D}$ is a dissection of **size** $n$. We say that $\mathscr{D}$ is a **regular** dissection if $a_j - a_{j-1} = (b - a)/n$ for all $j$, that is, if the points in the dissection are regularly spaced.

**Definition 11.2.** Let $f : [a, b] \to \mathbb{R}$ be a bounded function and $\mathscr{D}$ be a dissection of size $n$ of $[a, b]$. For each $j \in \{1, 2, \ldots, n\}$, let

$$m_j = \inf\{f(x) : a_{j-1} \le x \le a_j\},$$

$$\text{and} \quad M_j = \sup\{f(x) : a_{j-1} \le x \le a_j\}.$$

Note that these numbers exist, since $f$ is bounded. The **lower Riemann sum** of $f$ with respect to $\mathscr{D}$ is

$$l_{\mathscr{D}}(f) = \sum_{j=1}^{n} m_j(a_j - a_{j-1}),$$

and the **upper Riemann sum** of $f$ with respect to $\mathscr{D}$ is

$$u_{\mathscr{D}}(f) = \sum_{j=1}^{n} M_j(a_j - a_{j-1}).$$

The idea is that a dissection of size $n$ divides $[a, b]$ into $n$ subintervals, $[a_{j-1}, a_j]$ for $j = 1, 2, \ldots, n$. If $f(x) \ge 0$ for all $x$, then the lower Riemann sum can be visualized as the total area of the tallest rectangles with bases $[a_{j-1}, a_j]$ which fit under the graph $y = f(x)$ between $x = a$ and $x = b$. So $l_{\mathscr{D}}(f)$ is an underestimate of the area under the graph. Similarly, the upper Riemann sum can be visualized as the total area of the shortest rectangles with bases $[a_{j-1}, a_j]$ which the graph $y = f(x)$ between $x = a$ and $x = b$ fits under and is thus an overestimate of the area under the graph. This is illustrated in Figure 11.1.

**Example 11.3.** $\mathscr{D}_1 = \{0, 1\}$, $\mathscr{D}_2 = \{0, \frac{1}{2}, 1\}$, and $\mathscr{D}_3 = \{0, \frac{1}{2}, \frac{3}{4}, 1\}$ are dissections of $[0, 1]$. The function $f : [0, 1] \to \mathbb{R}$, $f(x) = x^2$ is bounded, and its lower and upper Riemann sums with respect to these dissections are:

$$l_{\mathscr{D}_1}(f) = 0(1 - 0) = 0$$

$$u_{\mathscr{D}_1}(f) = 1(1 - 0) = 1$$

$$l_{\mathscr{D}_2}(f) = 0(\frac{1}{2} - 0) + \frac{1}{4}(1 - \frac{1}{2}) = \frac{1}{8}$$

$$u_{\mathscr{D}_2}(f) = \frac{1}{4}(\frac{1}{2} - 0) + 1(1 - \frac{1}{2}) = \frac{5}{8}$$

$$l_{\mathscr{D}_3}(f) = 0(\frac{1}{2} - 0) + \frac{1}{4}(\frac{3}{4} - \frac{1}{2}) + \frac{9}{16}(1 - \frac{3}{4}) = \frac{13}{64}$$

$$u_{\mathscr{D}_3}(f) = \frac{1}{4}(\frac{1}{2} - 0) + \frac{9}{16}(\frac{3}{4} - \frac{1}{2}) + 1(1 - \frac{3}{4}) = \frac{33}{64}.$$

Note that

$$l_{\mathscr{D}_1} \le l_{\mathscr{D}_2} \le l_{\mathscr{D}_3} \le u_{\mathscr{D}_3} \le u_{\mathscr{D}_2} \le u_{\mathscr{D}_1}.$$

We will see shortly that this is no accident.

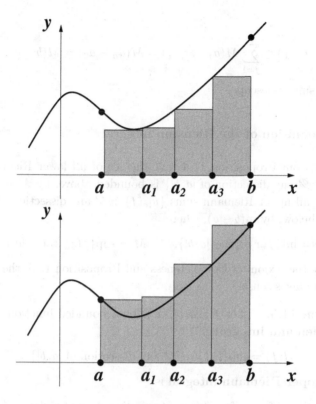

Fig. 11.1 Lower and upper Riemann sums of a function with respect to a dissection of size 4.

**Proposition 11.4.** *Let* $f : [a, b] \to \mathbb{R}$ *be bounded above by* $M$ *and below by* $m$, *and let* $\mathscr{D}$ *be any dissection of* $[a, b]$. *Then*

$$m(b - a) \leq l_{\mathscr{D}}(f) \leq u_{\mathscr{D}}(f) \leq M(b - a).$$

**Proof.** Let $\mathscr{D} = \{a_0, a_1, \ldots, a_n\}$ and $m_j$, $M_j$ be defined as in Definition 11.2. Then, for all $j$, $m \leq m_j \leq M_j \leq M$, so

$$l_{\mathscr{D}}(f) = \sum_{j=1}^{n} m_j(a_j - a_{j-1}) \leq \sum_{j=1}^{n} M_j(a_j - a_{j-1}) = u_{\mathscr{D}}(f)$$

and

$$l_{\mathscr{D}}(f) \geq \sum_{j=1}^{n} m(a_j - a_{j-1}) = m(a_n - a_0) = m(b - a),$$

and

$$u_{\mathscr{D}}(f) \le \sum_{j=1}^{n} M(a_j - a_{j-1}) = M(a_n - a_0) = M(b-a),$$

since the sums telescope. □

## 11.2　Definition of the Riemann integral

It follows from Proposition 11.4 that the set of all lower Riemann sums $\{l_{\mathscr{D}}(f) : \mathscr{D} \text{ any dissection of } [a,b]\}$ is bounded above, by $M(b-a)$, and the set of all upper Riemann sums $\{u_{\mathscr{D}}(f) : \mathscr{D} \text{ any dissection of } [a,b]\}$ is bounded below, by $m(b-a)$, where

$$m = \inf\{f(x) : x \in [a,b]\}, \qquad M = \sup\{f(x) : x \in [a,b]\}.$$

Hence, by the Axiom of Completeness and Proposition 1.23, the following definition makes sense.

**Definition 11.5.**　Let $f : [a,b] \to \mathbb{R}$ be a bounded function. Then its **lower Riemann integral** is

$$l(f) = \sup\{l_{\mathscr{D}}(f) : \mathscr{D} \text{ any dissection of } [a,b]\},$$

and its **upper Riemann integral** is

$$u(f) = \inf\{u_{\mathscr{D}}(f) : \mathscr{D} \text{ any dissection of } [a,b]\}.$$

We say that $f$ is **Riemann integrable** (on $[a,b]$) if $l(f) = u(f)$. In that case, we denote this commmon value by

$$\int_a^b f \quad \text{or} \quad \int_a^b f(x)dx,$$

and call this number the **Riemann integral** of $f$ (over $[a,b]$).

　　Geometrically, we can think of $l(f)$ as the least upper bound on the collection of all underestimates of the area under the curve $y = f(x)$, and $u(f)$ as the greatest lower bound on the collection of overestimates of the area under the curve. Then $\int_a^b f$ is, loosely, the unique number (where this exists) which is smaller than every overestimate and larger than every underestimate. As it stands, to compute this number, or even show that it exists, we need to consider the collection of all possible dissections of $[a,b]$. This is a very large and complicated set, so, before proceeding further, we need to develop some tools for handling $l(f)$ and $u(f)$ which allow us to avoid considering all possible dissections. The key idea is that of *refinement* of a dissection.

**Definition 11.6.** Given dissections $\mathcal{D}, \mathcal{D}'$ of $[a, b]$, we say that $\mathcal{D}'$ is a **refinement** of $\mathcal{D}$ if $\mathcal{D} \subseteq \mathcal{D}'$. If $\mathcal{D}' \backslash \mathcal{D}$ contains $k$ points, we say that $\mathcal{D}'$ is a $k$-**point refinement** of $\mathcal{D}$. Note that $\mathcal{D}$ is the unique 0-point refinement of itself.

The idea is that a refinement $\mathcal{D}'$ of $\mathcal{D}$ contains all the points in $\mathcal{D}$ and (unless $\mathcal{D}' = \mathcal{D}$) some more points too, so it splits $[a, b]$ up into more subintervals, at least some of which are narrower. Intuitively, one expects that passing from $\mathcal{D}$ to a refinement of $\mathcal{D}$ can only improve, that is, increase, the underestimate $l_{\mathcal{D}}(f)$. Similarly, passing to a refinement of $\mathcal{D}$, one expects, can only reduce the overestimate $u_{\mathcal{D}}(f)$. This expectation turns out to be essentially correct and is fundamental.

**Lemma 11.7 (Refinement Lemma).** *Let* $f : [a, b] \to \mathbb{R}$ *be bounded,* $\mathcal{D}, \mathcal{D}'$ *be dissections of* $[a, b]$, *and* $\mathcal{D}'$ *be a refinement of* $\mathcal{D}$. *Then*
$$l_{\mathcal{D}}(f) \le l_{\mathcal{D}'}(f) \le u_{\mathcal{D}'}(f) \le u_{\mathcal{D}}(f).$$

**Proof.** We prove this by induction on the size of $\mathcal{D}' \backslash \mathcal{D}$, that is, on the natural number $k$, where $\mathcal{D}'$ is a $k$-point refinement of $\mathcal{D}$.

First note that the result follows trivially from Proposition 11.4 if $k = 0$, since then $\mathcal{D}' = \mathcal{D}$.

Now assume that the result holds for every $k$-point refinement of every dissection $\mathcal{D}$ of $[a, b]$, and, for given $\mathcal{D}$, consider an arbitrary $(k+1)$-point refinement $\mathcal{D}'$ of $\mathcal{D}$. Then $\mathcal{D}'$ is a 1-point refinement of some $k$-point refinement $\widehat{\mathcal{D}}$ of $\mathcal{D}$ and, by assumption,
$$l_{\mathcal{D}}(f) \le l_{\widehat{\mathcal{D}}}(f) \le u_{\widehat{\mathcal{D}}}(f) \le u_{\mathcal{D}}(f).$$
It suffices, therefore, to show that
$$l_{\widehat{\mathcal{D}}}(f) \le l_{\mathcal{D}'}(f) \le u_{\mathcal{D}'}(f) \le u_{\widehat{\mathcal{D}}}(f). \tag{11.1}$$

Let $\widehat{\mathcal{D}} = \{a = a_0, a_1, \ldots, a_n = b\}$ and $\mathcal{D}' = \widehat{\mathcal{D}} \cup \{z\}$. Then there exists $k \in \{1, \ldots, n\}$ such that $z \in (a_{k-1}, a_k)$. Let
$$m_j = \inf\{f(x) : a_j \le x \le a_{j-1}\}$$
$$m' = \inf\{f(x) : a_{k-1} \le x \le z\}$$
$$m'' = \inf\{f(x) : z \le x \le a_k\}$$
$$M_j = \sup\{f(x) : a_j \le x \le a_{j-1}\}$$
$$M' = \sup\{f(x) : a_{k-1} \le x \le z\}$$
$$M'' = \sup\{f(x) : z \le x \le a_k\}$$

and note that, since $[a_{k-1}, z]$ and $[z, a_k]$ are subsets of $[a_{k-1}, a_k]$, we know immediately that $m', m'' \geq m_k$, and $M', M'' \leq M_k$. Now

$$l_{\mathscr{D}'}(f) = \sum_{j \in \{1,2,\ldots,n\} \setminus \{k\}} m_j(a_j - a_{j-1}) + m'(z - a_{k-1}) + m''(a_k - z)$$

$$= l_{\widehat{\mathscr{D}}}(f) - m_k(a_k - a_{k-1}) + m'(z - a_{k-1}) + m''(a_k - z)$$

$$\geq l_{\widehat{\mathscr{D}}}(f) - m_k(a_k - a_{k-1}) + m_k(z - a_{k-1}) + m_k(a_k - z)$$

$$= l_{\widehat{\mathscr{D}}},$$

and

$$u_{\mathscr{D}'}(f) = \sum_{j \in \{1,2,\ldots,n\} \setminus \{k\}} M_j(a_j - a_{j-1}) + M'(z - a_{k-1}) + M''(a_k - z)$$

$$= u_{\widehat{\mathscr{D}}}(f) - M_k(a_k - a_{k-1}) + M'(z - a_{k-1}) + M''(a_k - z)$$

$$\leq u_{\widehat{\mathscr{D}}}(f) - M_k(a_k - a_{k-1}) + M_k(z - a_{k-1}) + M_k(a_k - z)$$

$$= u_{\widehat{\mathscr{D}}}.$$

Hence, (11.1) follows from Proposition 11.4.                                 □

It follows immediately from the Refinement Lemma that *every* upper Riemann sum is at least as large as *every* lower Riemann sum, whatever (possibly different) dissections we use to compute them.

**Lemma 11.8.** *Let $\mathscr{D}, \mathscr{D}'$ be two dissections of $[a, b]$ and $f : [a, b] \to \mathbb{R}$ be bounded. Then $l_{\mathscr{D}}(f) \leq u_{\mathscr{D}'}(f)$.*

**Proof.** $\mathscr{D}'' = \mathscr{D} \cup \mathscr{D}'$ is a refinement of both $\mathscr{D}$ and $\mathscr{D}'$, so by the Refinement Lemma,

$$l_{\mathscr{D}}(f) \leq l_{\mathscr{D}''}(f) \leq u_{\mathscr{D}''}(f) \leq u_{\mathscr{D}'}(f).$$

□

From this, it follows that the upper Riemann integral is no less than the lower Riemann integral.

**Lemma 11.9.** *Let $f : [a, b] \to \mathbb{R}$ be bounded. Then $l(f) \leq u(f)$.*

**Proof.** Assume, towards a contradiction, that $l(f) > u(f)$. Then $l(f)$ is not a lower bound on the set of upper Riemann sums of $f$ (since $u(f)$ is, by definition, the *greatest* lower bound on this set). Hence, there exists a dissection $\mathscr{D}$ such that $u_{\mathscr{D}}(f) < l(f)$. Hence, $u_{\mathscr{D}}(f)$ is not an upper bound on the set of lower Riemann sums of $f$ (since $l(f)$ is, by definition, the *least* upper bound on this set), so there exists $\mathscr{D}'$ such that $l_{\mathscr{D}'}(f) > u_{\mathscr{D}}(f)$. But this contradicts Lemma 11.8.                                 □

We are now, finally, in a position to compute some nontrivial Riemann integrals.

**Example 11.10.** Let $f : [0, 2] \to \mathbb{R}$ be the "step" function

$$f(x) = \begin{cases} 1 & \text{if } 0 \le x < 1, \\ 2 & \text{if } 1 \le x \le 2. \end{cases}$$

Show that $f$ is Riemann integrable on $[0, 2]$ and compute $\int_0^2 f$.

*Solution:* For each $r \in (0, 1)$ define the dissection $\mathscr{D}_r = \{0, 1 - r, 1 + r, 2\}$ of $\mathscr{D}_r$ (see Figure 11.2). Then the lower and upper Riemann sums of $f$ with respect to this dissection are

$$l_{\mathscr{D}_r}(f) = 1(1 - r - 0) + 1((1 + r) - (1 - r)) + 2(2 - (1 + r))$$
$$= 3 - r$$
$$u_{\mathscr{D}_r}(f) = 1(1 - r - 0) + 2((1 + r) - (1 - r)) + 2(2 - (1 + r))$$
$$= 3 + r.$$

Hence, $l(f)$ is the supremum of a set (the set of all lower sums) which contains $\{3 - r \; : \; 0 < r < 1\} = (2, 3)$, so $l(f) \ge 3$. Similarly, $u(f)$ is the infimum of a set (the set of all upper sums) which contains $\{3 + r : 0 < r < 1\} = (3, 4)$, so $u(f) \le 3$. Hence $l(f) \ge u(f)$. But by Lemma 11.9, $l(f) \le u(f)$, so $l(f) = u(f)$, that is, $f$ is Riemann integrable. Furthermore

$$\int_0^2 f = l(f) \ge 3 \quad \text{and} \quad \int_0^2 f = u(f) \le 3$$

so we conclude that $\int_0^2 f = 3$. $\qquad\qquad\square$

**Example 11.11.** Let $f : [0, 1] \to \mathbb{R}$ be the function $f(x) = x^2$. Show that $f$ is Riemann integrable on $[0, 1]$ and compute $\int_0^1 f$.

*Solution:* For each integer $n \ge 1$ let $\mathscr{D}_n$ be the *regular* dissection of $[0, 1]$ of size $n$, that is, the dissection that divides $[0, 1]$ into $n$ subintervals of equal width,

$$\mathscr{D}_n = \left\{ 0, \frac{1}{n}, \frac{2}{n}, \dots, \frac{n-1}{n}, 1 \right\}.$$

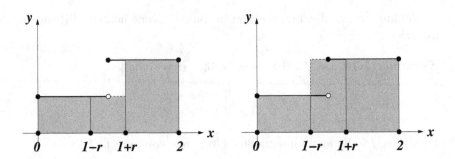

Fig. 11.2   The family of dissections, $\mathscr{D}_r$, $0 < r < 1$, used in Example 11.10.

Each subinterval $[a_{j-1}, a_j] = [(j-1)/n, j/n]$ has width $1/n$, and, since the function $f$ is increasing,

$$m_j = \inf\left\{ f(x) \;:\; \frac{j-1}{n} \le x \le \frac{j}{n} \right\} = f\left(\frac{j-1}{n}\right) = \frac{(j-1)^2}{n^2},$$

$$M_j = \sup\left\{ f(x) \;:\; \frac{j-1}{n} \le x \le \frac{j}{n} \right\} = f\left(\frac{j}{n}\right) = \frac{j^2}{n^2},$$

so the lower and upper Riemann sums with respect to $\mathscr{D}_n$ are

$$l_{\mathscr{D}_n}(f) = \frac{1}{n} \sum_{j=1}^{n} \frac{(j-1)^2}{n^2} = \frac{1}{n^3} \sum_{k=1}^{n-1} k^2,$$

$$u_{\mathscr{D}_n}(f) = \frac{1}{n} \sum_{j=1}^{n} \frac{j^2}{n^2} = \frac{1}{n^3} \sum_{j=1}^{n} j^2.$$

I claim that, for all $n \in \mathbb{Z}^+$,

$$\sum_{j=1}^{n} j^2 = \frac{1}{6} n(n+1)(2n+1),$$

and leave the proof of this as an exercise (hint: use induction!). Hence, for each $n \in \mathbb{Z}^+$,

$$l_{\mathscr{D}_n}(f) = \frac{1}{6}\left(1 - \frac{1}{n}\right)\left(2 - \frac{1}{n}\right),$$

$$u_{\mathscr{D}_n}(f) = \frac{1}{6}\left(1 + \frac{1}{n}\right)\left(2 + \frac{1}{n}\right).$$

By definition, $l(f) \ge l_{\mathscr{D}_n}(f)$ for all $n$ (it is an upper bound on a set containing every $l_{\mathscr{D}_n}(f)$), and $l_{\mathscr{D}_n}(f) \to \frac{1}{3}$. Hence $l(f) \ge \frac{1}{3}$ (Proposition 3.4).

Similarly, $u(f) \leq u_{\mathcal{D}_n}(f)$ for all $n$ (it is a lower bound on a set containing every $u_{\mathcal{D}_n}(f)$), and $u_{\mathcal{D}_n}(f) \to \frac{1}{3}$. Hence $u(f) \leq \frac{1}{3}$ (Proposition 3.4).

But, by Lemma 11.9, $l(f) \leq u(f)$, so we conclude that $l(f) = u(f)$, that is, $f$ is Riemann integrable. Furthermore,

$$\int_0^1 f = l(f) \geq \frac{1}{3} \quad \text{and} \quad \int_0^1 f = u(f) \leq \frac{1}{3},$$

whence $\int_0^1 f = \frac{1}{3}$. $\qquad\square$

It is convenient to extend the definition of dissection to include the case where the interval is $[a, a] = \{a\}$. The only dissection of $[a, a]$ is the singleton set $\mathcal{D} = \{a\}$ (a dissection of size 0). Every function $f : [a, a] \to \mathbb{R}$ is bounded, above and below by $f(a)$, so the one and only lower Riemann sum is

$$l_{\{a\}}(f) = f(a)(a - a) = 0,$$

and the one and only upper Riemann sum is

$$u_{\{a\}}(f) = f(a)(a - a) = 0.$$

It follows that every function is integrable on $[a, a]$, and

$$\int_a^a f = 0.$$

Given a Riemann integrable function $f$ on $[a, b]$ where $a \leq b$, it is also convenient to define

$$\int_b^a f = -\int_a^b f.$$

I leave it to the reader to verify that all the results we prove about $\int_a^b f$ trivially extend to the case $a \geq b$ with these conventions.

## 11.3 A sequential characterization of integrability

We can generalize the argument used in Example 11.11 to give a rather elegant (and useful) characterization of Riemann integrability in terms of *sequences* of dissections. This fits nicely with our philosophy of formulating all the fundamental notions of real analysis in terms of sequences.

**Theorem 11.12.** *Let $f : [a, b] \to \mathbb{R}$ be a bounded function. Then $f$ is Riemann integrable if and only if there exists a sequence $(\mathcal{D}_n)$ of dissections of $[a, b]$ such that $u_{\mathcal{D}_n}(f) - l_{\mathcal{D}_n}(f) \to 0$. In this case,*

$$\int_a^b f = \lim l_{\mathcal{D}_n}(f) = \lim u_{\mathcal{D}_n}(f).$$

**Proof.** Let $\mathcal{D}$ be the set of all dissections of $[a,b]$, $\mathcal{L} = \{l_\mathscr{D}(f) : \mathscr{D} \in \mathcal{D}\}$ the set of all lower Riemann sums of $f$, and $\mathcal{U} = \{u_\mathscr{D}(f) : \mathscr{D} \in \mathcal{D}\}$ the set of all upper Riemann sums of $f$, so $l(f) = \sup \mathcal{L}$ and $u(f) = \inf \mathcal{U}$.

We first prove the "if" direction. So, assume that a sequence of dissections $\mathscr{D}_n \in \mathcal{D}$ exists such that $u_{\mathscr{D}_n}(f) - l_{\mathscr{D}_n}(f) \to 0$. By Lemma 11.8, every $u_{\mathscr{D}_n}(f)$ is an upper bound on $\mathcal{L}$. Now, $l(f)$ is the *least* upper bound on $\mathcal{L}$, so for all $n \in \mathbb{Z}^+$, $l(f) \leq u_{\mathscr{D}_n}(f)$. Hence, for all $n \in \mathbb{Z}^+$,

$$0 \leq l(f) - l_{\mathscr{D}_n}(f) \leq u_{\mathscr{D}_n}(f) - l_{\mathscr{D}_n}(f),$$

and hence $l_{\mathscr{D}_n}(f) \to l(f)$ by the Squeeze Rule. Similarly, by Lemma 11.8, every $l_{\mathscr{D}_n}(f)$ is a lower bound on $\mathcal{U}$, and $u(f)$ is the *greatest* lower bound on $\mathcal{U}$, so $u(f) \geq l_{\mathscr{D}_n}(f)$. Hence, for all $n \in \mathbb{Z}^+$,

$$0 \leq u_{\mathscr{D}_n}(f) - u(f) \leq u_{\mathscr{D}_n}(f) - l_{\mathscr{D}_n}(f),$$

so $u_{\mathscr{D}_n}(f) \to u(f)$ by the Squeeze Rule. Hence

$$u_{\mathscr{D}_n}(f) - l_{\mathscr{D}_n}(f) \to u(f) - l(f)$$

(Algebra of Limits). But this sequence converges to 0, and limits are unique, so $u(f) = l(f)$. Hence, $f$ is Riemann integrable on $[a,b]$, and

$$\int_a^b f = l(f) = \lim l_{\mathscr{D}_n}(f) = u(f) = \lim u_{\mathscr{D}_n}(f).$$

We now prove the "only if" direction. So assume that $f$ is Riemann integrable on $[a,b]$, that is, $l(f) = u(f)$. By Proposition 3.16, there exist sequences of dissections $\mathscr{D}_n' \in \mathcal{D}$ and $\mathscr{D}_n'' \in \mathcal{D}$ such that $l_{\mathscr{D}_n'}(f) \to l(f)$ and $u_{\mathscr{D}_n''}(f) \to u(f)$. Let $\mathscr{D}_n = \mathscr{D}_n' \cup \mathscr{D}_n''$, and note that this is a refinement of both $\mathscr{D}_n'$ and $\mathscr{D}_n''$ so, by the Refinement Lemma,

$$l_{\mathscr{D}_n'}(f) \leq l_{\mathscr{D}_n}(f) \leq l(f) \qquad \text{and} \qquad u(f) \leq u_{\mathscr{D}_n}(f) \leq u_{\mathscr{D}_n''}(g).$$

Hence, by the Squeeze Rule, $l_{\mathscr{D}_n}(f) \to l(f)$ and $u_{\mathscr{D}_n}(f) \to u(f)$. But $l(f) = u(f)$, so $u_{\mathscr{D}_n}(f) - l_{\mathscr{D}_n}(f) \to 0$ (Algebra of Limits), and

$$\int_a^b f = l(f) = \lim l_{\mathscr{D}_n}(f) = u(f) = \lim u_{\mathscr{D}_n}(f).$$

$\square$

To illustrate the power of Theorem 11.12, let's use it to prove that *every* monotonic (i.e. increasing or decreasing) function is Riemann integrable.

**Theorem 11.13.** *Let $f : [a,b] \to \mathbb{R}$ be monotonic. Then $f$ is Riemann integrable.*

**Proof.** First note that if $f$ is monotonic, it is certainly bounded (below by $f(a)$ and above by $f(b)$ if it is increasing, above by $f(a)$ and below by $f(b)$ if it is decreasing), so $l(f)$ and $u(f)$ exist. For each $n \in \mathbb{Z}^+$ let $\mathscr{D}_n$ be the regular dissection of $[a, b]$ of size $n$, that is, the dissection which divides $[a, b]$ into $n$ subintervals $[a_{j-1}, a_j]$ each of width $(b - a)/n$. We handle the cases where $f$ is increasing and $f$ is decreasing separately.

So, assume that $f$ is increasing. Then, for each $j = 1, \ldots, n$,

$$m_j = \inf\{f(x) : a_{j-1} \leq x \leq a_j\} = f(a_{j-1})$$
$$M_j = \sup\{f(x) : a_{j-1} \leq x \leq a_j\} = f(a_j),$$

and so

$$u_{\mathscr{D}_n}(f) - l_{\mathscr{D}_n}(f) = \frac{b - a}{n} \sum_{j=1}^{n} (f(a_j) - f(a_{j-1}))$$
$$= \frac{b - a}{n}(f(b) - f(a))$$

since the sum telescopes. Hence $u_{\mathscr{D}_n}(f) - l_{\mathscr{D}_n}(f) \to 0$, so $f$ is Riemann integrable on $[a, b]$ by Theorem 11.12.

Finally, assume that $f$ is decreasing. Then $m_j = f(a_j)$ and $M_j = f(a_{j-1})$ so

$$u_{\mathscr{D}_n}(f) - l_{\mathscr{D}_n}(f) = \frac{b - a}{n} \sum_{j=1}^{n} (f(a_{j-1}) - f(a_j)) = \frac{b - a}{n}(f(a) - f(b)) \to 0.$$

Hence $f$ is Riemann integrable on $[a, b]$ by Theorem 11.12. $\square$

Note that Theorem 11.13 partially covers both Examples 11.10 and 11.11: in both these examples the function in question is increasing so Theorem 11.13 implies that they're both Riemann integrable. It does not, however, tell us anything about the *value* of $\int_a^b f$, only that it exists. Note also that the function in Example 11.10 is discontinuous, at a single point ($x = 1$), which illustrates that while a discontinuous function cannot be differentiable (Proposition 9.6), it certainly can be integrable. In fact, we can construct examples of functions which are discontinuous at infinitely many points in $[a, b]$ and yet are still Riemann integrable on $[a, b]$.

**Example 11.14.** Consider the function $f : [0, 1] \to \mathbb{R}$ defined so that, $f(0) = 0$ and, for all $x \in (\frac{1}{n+1}, \frac{1}{n}]$, where $n \in \mathbb{Z}^+$, $f(x) = \frac{1}{n}$ (see Figure 11.3). By construction, $f$ is increasing and so is Riemann integrable on $[0, 1]$ by Theorem 11.13. Note, however, that $f$ is discontinuous at every point $\frac{1}{n}$ for $n \geq 2$.

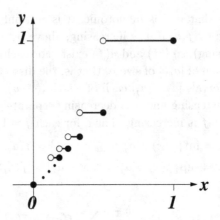

Fig. 11.3  The highly discontinuous function considered in Example 11.14. Note that this function is increasing, and hence is Riemann integrable by Theorem 11.13.

If even such an extreme example as this is Riemann integrable, you might wonder whether there are any bounded functions which *fail* to be integrable. Fear not: such functions certainly do exist. Here is an example.

**Example 11.15.** Let $f : [0,1] \to \mathbb{R}$ such that $f(x) = 0$ if $x \in \mathbb{Q}$ and $f(x) = 1$ if $x \notin \mathbb{Q}$. I claim that $f$ is not Riemann integrable.

**Proof.** Let $\mathscr{D} = \{a_0, \ldots, a_n\}$ be any dissection of $[0,1]$. Then, by the Density Theorems of $\mathbb{Q}$ in $\mathbb{R}$ and $\mathbb{R}\backslash\mathbb{Q}$ in $\mathbb{R}$ (Theorems 1.25 and 1.34), every subinterval $[a_{j-1}, a_j]$ contains both rational and irrational members, so $m_j = 0$ and $M_j = 1$ for all $j$. Hence

$$l_{\mathscr{D}}(f) = \sum_{j=1}^{n} 0(a_j - a_{j-1}) = 0$$

$$u_{\mathscr{D}}(f) = \sum_{j=1}^{n} 1(a_j - a_{j-1}) = a_n - a_0 = 1.$$

Since this is true for *all* dissections of $[0,1]$, $l(f) = \sup\{0\} = 0$ and $u(f) = \inf\{1\} = 1$, so $l(f) \neq u(f)$, that is, $f$ is not Riemann integrable.  $\square$

Theorem 11.13 gives us one interesting class of functions $f : [a,b] \to \mathbb{R}$ that are Riemann integrable: those that are monotonic. Our next theorem gives us another rather more useful class: continuous functions.

**Theorem 11.16.**  *Let $f : [a,b] \to \mathbb{R}$ be continuous. Then $f$ is Riemann integrable.*

**Proof.** First note that, since $f$ is continuous, it is certainly bounded (by the Extreme Value Theorem), so $l(f)$ and $u(f)$ exist. Assume, towards a contradiction, that $l(f) \neq u(f)$ and hence, by Lemma 11.9, $u(f) > l(f)$. Let $\varepsilon = u(f) - l(f) > 0$.

Consider the sequence $(\mathscr{D}_n)$ of regular dissections of $[a, b]$ of size $2^n$. Then $\mathscr{D}_n \subset \mathscr{D}_{n+1}$ for all $n$, so by the Refinement Lemma, $l_{\mathscr{D}_n}(f)$ is increasing and $u_{\mathscr{D}_n}(f)$ is decreasing. Hence $l_{\mathscr{D}_n}(f) \to K \leq l(f)$ and $u_{\mathscr{D}_n}(f) \to L \geq u(f)$ for some numbers $K$ and $L$ (by the Monotone Convergence Theorem and Proposition 3.4), and

$$u_{\mathscr{D}_n}(f) - l_{\mathscr{D}_n}(f) \to L - K \geq u(f) - l(f) = \varepsilon.$$

As usual, let

$$m_j = \inf\{f(x) : a_{j-1} \leq x \leq a_j\}$$
$$\text{and} \quad M_j = \sup\{f(x) : a_{j-1} \leq x \leq a_j\}.$$

Then

$$u_{\mathscr{D}_n}(f) - l_{\mathscr{D}_n}(f) = \frac{b-a}{2^n} \sum_{j=1}^{2^n}(M_j - m_j) \geq \varepsilon.$$

This sum consists of $2^n$ terms, each non-negative, so at least one term must be greater than or equal to $\varepsilon/2^n$. That is, there must exist $j \in \{1, 2, \ldots, 2^n\}$ such that

$$M_j - m_j \geq \frac{\varepsilon}{b-a}.$$

But $f : [a_{j-1}, a_j] \to \mathbb{R}$ is continuous so, by the Extreme Value Theorem, $f$ attains both a maximum and minimum value on $[a_{j-1}, a_j]$, that is, there are points, $x_n$ and $y_n$ say, in $[a_{j-1}, a_j]$ such that $f(x_n) = M_j$ and $f(y_n) = m_j$. Hence, for each $n \in \mathbb{Z}^+$ there exist $x_n, y_n \in [a, b]$ such that

$$|x_n - y_n| \leq \frac{b-a}{2^n}, \tag{11.2}$$

$$f(x_n) - f(y_n) \geq \frac{\varepsilon}{b-a}. \tag{11.3}$$

Consider the sequence $(x_n)$. It is bounded, so, by the Bolzano–Weierstrass Theorem, it has a convergent subsequence $x_{n_k} \to c \in [a, b]$. By (11.2),

$$x_{n_k} - \frac{b-a}{2^{n_k}} \leq y_{n_k} \leq x_{n_k} + \frac{b-a}{2^{n_k}}$$

so $y_{n_k} \to c$ also, by the Squeeze Rule. Now $f$ is continuous, so $f(x_{n_k}) \to f(c)$ and $f(y_{n_k}) \to f(c)$, and hence $f(x_{n_k}) - f(y_{n_k}) \to f(c) - f(c) = 0$. But this contradicts (11.3) and Proposition 3.4. $\square$

## 11.4   Elementary properties of the Riemann integral

Theorem 11.12 is a convenient tool for establishing many useful properties of the Riemann integral.

**Theorem 11.17.** *Let $f$ be Riemann integrable on $[a,b]$ and on $[b,c]$. Then $f$ is Riemann integrable on $[a,c]$ and*

$$\int_a^c f = \int_a^b f + \int_b^c f.$$

**Proof.** By Theorem 11.12, there exist sequences of dissections $(\mathscr{D}_n')$ of $[a,b]$ and $(\mathscr{D}_n'')$ of $[b,c]$ such that $l_{\mathscr{D}_n'}(f) \to \int_a^b f$, $u_{\mathscr{D}_n'}(f) \to \int_a^b f$, $l_{\mathscr{D}_n''}(f) \to \int_b^c f$, and $u_{\mathscr{D}_n''}(f) \to \int_b^c f$. Let $\mathscr{D}_n = \mathscr{D}_n' \cup \mathscr{D}_n''$. Then $\mathscr{D}_n$ is a dissection of $[a,c]$ and it follows directly from Definition 11.2 that $l_{\mathscr{D}_n}(f) = l_{\mathscr{D}_n'}(f) + l_{\mathscr{D}_n''}(f)$ and $u_{\mathscr{D}_n}(f) = u_{\mathscr{D}_n'}(f) + u_{\mathscr{D}_n''}(f)$. Hence, by the Algebra of Limits, $l_{\mathscr{D}_n}(f) \to \int_a^b f + \int_b^c f$ and $u_{\mathscr{D}_n}(f) \to \int_a^b f + \int_b^c f$, so the claim follows from Theorem 11.12.          □

The converse of Theorem 11.17 is also true:

**Theorem 11.18.** *Let $f$ be Riemann integrable on $[a,c]$ and $b \in [a,c]$. Then $f$ is Riemann integrable on $[a,b]$ and $[b,c]$, and*

$$\int_a^b f + \int_b^c f = \int_a^c f.$$

**Proof.** By Theorem 11.12, there exists a sequence of dissections $(\mathscr{D}_n)$ of $[a,c]$ such that $u_{\mathscr{D}_n}(f) - l_{\mathscr{D}_n}(f) \to 0$. Let $\mathscr{D}_n' = \mathscr{D}_n \cup \{b\}$. This is a refinement of $\mathscr{D}_n$, so by the Refinement Lemma and the Squeeze Rule $u_{\mathscr{D}_n'}(f) - l_{\mathscr{D}_n'}(f) \to 0$ also. Let $\mathscr{D}_n^L = \mathscr{D}_n' \cap [a,b]$ and $\mathscr{D}_n^R = \mathscr{D}_n' \cap [b,c]$. Then $\mathscr{D}_n^L$ is a dissection of $[a,b]$ and $\mathscr{D}_n^R$ is a dissection of $[b,c]$, and $\mathscr{D}_n' = \mathscr{D}_n^L \cup \mathscr{D}_n^R$. Now

$$l_{\mathscr{D}_n'}(f) = l_{\mathscr{D}_n^L}(f) + l_{\mathscr{D}_n^R}(f), \quad \text{and} \quad u_{\mathscr{D}_n'}(f) = u_{\mathscr{D}_n^L}(f) + u_{\mathscr{D}_n^R}(f),$$

so

$$u_{\mathscr{D}_n'}(f) - l_{\mathscr{D}_n'}(f) = [u_{\mathscr{D}_n^L}(f) - l_{\mathscr{D}_n^L}(f)] + [u_{\mathscr{D}_n^R}(f) - l_{\mathscr{D}_n^R}(f)].$$

This is a sum of non-negative sequences, so

$$0 \le u_{\mathscr{D}_n^L}(f) - l_{\mathscr{D}_n^L}(f) \le u_{\mathscr{D}_n'}(f) - l_{\mathscr{D}_n'}(f)$$
$$0 \le u_{\mathscr{D}_n^R}(f) - l_{\mathscr{D}_n^R}(f) \le u_{\mathscr{D}_n'}(f) - l_{\mathscr{D}_n'}(f).$$

Hence, by the Squeeze Rule, $u_{\mathscr{D}_n^L}(f) - l_{\mathscr{D}_n^L}(f) \to 0$ and $u_{\mathscr{D}_n^R}(f) - l_{\mathscr{D}_n^R}(f) \to 0$, so $f$ is Riemann integrable on $[a, b]$ and $[b, c]$ by Theorem 11.12. The fact that

$$\int_a^b f + \int_b^c f = \int_a^c f.$$

then follows from Theorem 11.17. $\qquad\square$

**Corollary 11.19.** *Let $f$ be Riemann integrable on $[a, b]$ and $[c, d] \subseteq [a, b]$. Then $f$ is Riemann integrable on $[c, d]$.*

**Proof.** $c \in [a, b]$, so $f$ is integrable on $[c, b]$ by Theorem 11.18. Similarly, $d \in [c, b]$, so $f$ is integrable on $[c, d]$ by Theorem 11.18. $\qquad\square$

We next want to prove that the Riemann integral is *linear*, that is, $\int_a^b (\alpha f + \beta g) = \alpha \int_a^b f + \beta \int_a^b g$, for any constants $\alpha, \beta$. To do this, we need the following:

**Lemma 11.20.** *Let $f, g : [a, b] \to \mathbb{R}$ be bounded functions and $\mathscr{D}$ be a dissection of $[a, b]$. Then*

$$l_{\mathscr{D}}(f + g) \geq l_{\mathscr{D}}(f) + l_{\mathscr{D}}(g) \quad and \quad u_{\mathscr{D}}(f + g) \leq u_{\mathscr{D}}(f) + u_{\mathscr{D}}(g).$$

**Proof.** Let $\mathscr{D} = \{a_0, a_1, \ldots, a_n\}$ and $I_j = [a_{j-1}, a_j]$. Let

$$m_j(f) = \inf\{f(x) : x \in I_j\},$$
$$M_j(f) = \sup\{f(x) : x \in I_j\},$$

and $m_j(g), M_j(g), m_j(g + f)$, and $M_j(g + f)$ be defined similarly. Then, for all $x \in I_j$, $f(x) + g(x) \geq m_j(f) + m_j(g)$, so $m_j(f) + m_j(g)$ is certainly a lower bound on $\{f(x) + g(x) : x \in I_j\}$. Since $m_j(f + g)$ is the *greatest* lower bound on this set, it follows that $m_j(f + g) \geq m_j(f) + m_j(g)$. Hence

$$l_{\mathscr{D}}(f + g) = \sum_{j=1}^n m_j(f + g)(a_j - a_{j-1})$$

$$\geq \sum_{j=1}^n (m_j(f) + m_j(g))(a_j - a_{j-1}) = l_{\mathscr{D}}(f) + l_{\mathscr{D}}(g).$$

Similarly, for all $x \in I_j$, $f(x) + g(x) \leq M_j(f) + M_j(g)$, so $M_j(f) + M_j(g)$ is certainly an upper bound on $\{f(x) + g(x) : x \in I_j\}$. Since $M_j(f + g)$ is the *least* upper bound on this set, it follows that $M_j(f + g) \le M_j(f) + M_j(g)$. Hence

$$u_{\mathscr{D}}(f + g) = \sum_{j=1}^{n} M_j(f + g)(a_j - a_{j-1})$$

$$\le \sum_{j=1}^{n} (M_j(f) + M_j(g))(a_j - a_{j-1}) = u_{\mathscr{D}}(f) + u_{\mathscr{D}}(g).$$

$\square$

**Theorem 11.21 (Linearity of the Riemann Integral).** *Let* $f, g$ *be Riemann integrable on* $[a, b]$, *and* $\alpha \in \mathbb{R}$ *be a constant. Then*

*(i)* $\alpha f$ *is Riemann integrable on* $[a, b]$, *and* $\int_a^b \alpha f = \alpha \int_a^b f$,

*(ii)* $f + g$ *is Riemann integrable on* $[a, b]$, *and* $\int_a^b (f + g) = \int_a^b f + \int_a^b g$.

**Proof.** (i) By Theorem 11.12, there is a sequence of dissections $(\mathscr{D}_n)$ such that $l_{\mathscr{D}_n}(f) \to \int_a^b f$ and $u_{\mathscr{D}_n}(f) \to \int_a^b f$. It follows directly from Definition 11.2 that $l_{\mathscr{D}_n}(\alpha f) = \alpha l_{\mathscr{D}_n}(f)$ and $u_{\mathscr{D}_n}(\alpha f) = \alpha u_{\mathscr{D}_n}(f)$ if $\alpha \ge 0$, and $l_{\mathscr{D}_n}(\alpha f) = \alpha u_{\mathscr{D}_n}(f)$ and $u_{\mathscr{D}_n}(\alpha f) = \alpha l_{\mathscr{D}_n}(f)$ if $\alpha < 0$. In either case, $l_{\mathscr{D}_n}(\alpha f) \to \alpha \int_a^b f$ and $u_{\mathscr{D}_n}(\alpha f) \to \alpha \int_a^b f$ by the Algebra of Limits, and the claim follows from Theorem 11.12.

(ii) Let $(\mathscr{D}_n)$ be as defined above. By Theorem 11.12, there is also a sequence of dissections $(\mathscr{D}_n')$ such that $l_{\mathscr{D}_n'}(g) \to \int_a^b g$ and $u_{\mathscr{D}_n'}(g) \to \int_a^b g$. Let $\mathscr{D}_n'' = \mathscr{D}_n \cup \mathscr{D}_n'$. This is a refinement of both $\mathscr{D}_n$ and $\mathscr{D}_n'$ and so, by the Refinement Lemma

$$l_{\mathscr{D}_n}(f) \le l_{\mathscr{D}_n''}(f) \le \int_a^b f,$$

$$l_{\mathscr{D}_n'}(g) \le l_{\mathscr{D}_n''}(g) \le \int_a^b g,$$

$$\int_a^b f \le u_{\mathscr{D}_n''}(f) \le u_{\mathscr{D}_n}(f),$$

$$\int_a^b f \le u_{\mathscr{D}_n''}(g) \le u_{\mathscr{D}_n'}(g).$$

Hence, by the Squeeze Rule, $l_{\mathscr{D}_n''}(f) \to \int_a^b f$, $l_{\mathscr{D}_n''}(g) \to \int_a^b g$, $u_{\mathscr{D}_n''}(f) \to \int_a^b f$ and $u_{\mathscr{D}_n''}(f) \to \int_a^b g$. Now, by Lemma 11.20,

$$l_{\mathscr{D}_n''}(f) + l_{\mathscr{D}_n''}(g) \le l_{\mathscr{D}_n''}(f + g) \le l(f + g),$$

and

$$u(f+g) \leq u_{\mathscr{D}_n''}(f+g) \leq u_{\mathscr{D}_n''}(f) + u_{\mathscr{D}_n''}(g).$$

But both $l_{\mathscr{D}_n''}(f) + l_{\mathscr{D}_n''}(g)$ and $u_{\mathscr{D}_n''}(f) + u_{\mathscr{D}_n''}(g)$ converge to $\int_a^b f + \int_a^b g$ so, by Proposition 3.4, $\int_a^b f + \int_a^b g \leq l(f+g)$ and $u(f+g) \leq \int_a^b f + \int_a^b g$. It follows that $l(f+g) \geq u(f+g)$ and hence, by Lemma 11.9, that $l(f+g) = u(f+g)$, that is, $f+g$ is Riemann integrable. Furthermore

$$\int_a^b (f+g) = l(f+g) \geq \int_a^b f + \int_a^b g$$

and

$$\int_a^b (f+g) = u(f+g) \leq \int_a^b f + \int_a^b g,$$

so $\int_a^b(f+g) = \int_a^b f + \int_a^b g$, as claimed. □

So we know that the sum of two integrable functions is integrable. What about the *product* of two integrable functions? Again, this turns out to be integrable, but the proof is not so straightforward.

**Lemma 11.22.** *Let* $f : [a,b] \to \mathbb{R}$ *be bounded above by* $K \geq 0$ *and below by* $-K$, *and* $\mathscr{D}$ *be a dissection of* $[a,b]$. *Then*

$$u_{\mathscr{D}}(f^2) - l_{\mathscr{D}}(f^2) \leq 2K(u_{\mathscr{D}}(f) - l_{\mathscr{D}}(f)).$$

**Proof.** We first note that, for all $x, y \in [a,b]$,

$$f(x)^2 - f(y)^2 \leq |f(x)^2 - f(y)^2|$$
$$= |f(x) - f(y)||f(x) + f(y)|$$
$$\leq 2K|f(x) - f(y)|. \tag{11.4}$$

Now, let $\mathscr{D} = \{a_0, a_1, \ldots, a_n\}$ be the dissection of $[a,b]$, $I_j = [a_{j-1}, a_j]$, $|I_j| = a_j - a_{j-1}$,

$$A_j = \{f(x) : x \in I_j\}, \qquad B_j = \{f(x)^2 : x \in I_j\},$$

and define

$$M_j = \sup A_j, \qquad m_j = \inf A_j,$$
$$\widehat{M}_j = \sup B_j, \qquad \widehat{m}_j = \inf B_j.$$

Then

$$u_{\mathscr{D}}(f) - l_{\mathscr{D}}(f) = \sum_{j=1}^n (M_j - m_j)|I_j|,$$

and

$$u_{\mathscr{D}}(f^2) - l_{\mathscr{D}}(f^2) = \sum_{j=1}^n (\widehat{M}_j - \widehat{m}_j)|I_j|.$$

Now, recall that, given a bounded set $A \subset \mathbb{R}$,

$$\sup A - \inf A = \sup \operatorname{diff} A,$$

where $\operatorname{diff} A := \{x - y \,:\, x, y \in A\}$ (see Lemma 1.24 on page 12). Hence

$$
\begin{aligned}
\widehat{M}_j - \widehat{m}_j &= \sup B_j - \inf B_j \\
&= \sup \operatorname{diff} B_j \\
&= \sup \{ f(x)^2 - f(y)^2 \,:\, x, y \in I_j \} \\
&\leq \sup \{ 2K |f(x) - f(y)| \,:\, x, y \in I_j \} \qquad \text{by (11.4)} \\
&\leq \sup \{ 2K (f(x) - f(y)) \,:\, x, y \in I_j \} \qquad \begin{array}{l}\text{since the 1st set is a}\\ \text{subset of the second}\end{array} \\
&= 2K \sup \operatorname{diff} A_j \\
&= 2K (\sup A_j - \inf A_j) \qquad \text{by Lemma 1.24} \\
&= 2K (M_j - m_j).
\end{aligned}
$$

This holds for all $j \in \{1, \ldots, n\}$, so the claim immediately follows. $\qquad\square$

**Theorem 11.23.** *Let* $f : [a, b] \to \mathbb{R}$ *be bounded and Riemann integrable on* $[a, b]$. *Then* $f^2 : [a, b] \to \mathbb{R}$ *is bounded and Riemann integrable on* $[a, b]$.

**Proof.** Certainly, if $f$ is bounded, above by $K$, below by $-K$ say, then $f^2$ is bounded (above by $K^2$, below by $0$). Since $f$ is Riemann integrable, there exists a sequence $(\mathscr{D}_n)$ of dissections of $[a, b]$ such that $u_{\mathscr{D}_n}(f) - l_{\mathscr{D}_n}(f) \to 0$ (Theorem 11.12). But then, by Lemma 11.22,

$$0 \leq u_{\mathscr{D}_n}(f^2) - l_{\mathscr{D}_n}(f^2) \leq 2K(u_{\mathscr{D}_n}(f) - l_{\mathscr{D}_n}(f)),$$

so $u_{\mathscr{D}_n}(f^2) - l_{\mathscr{D}_n}(f^2) \to 0$ by the Squeeze Rule. Hence, $f^2$ is Riemann integrable (Theorem 11.12). $\qquad\square$

## Corollary 11.24 (Algebra Property of Integrable Functions).

*Let* $f, g$ *be bounded functions Riemann integrable on* $[a, b]$ *and* $\alpha \in \mathbb{R}$ *be a constant. Then*

$$\alpha f, \quad f + g, \quad and \quad fg$$

*are all Riemann integrable on* $[a, b]$.

**Proof.** We have already proved that $\alpha f$ and $f + g$ are integrable (Theorem 11.21) so it remains to show that $fg$ is integrable. By Theorem 11.23, $f^2$, $g^2$ and $(f + g)^2$ are all integrable. Hence, by Theorem 11.21,

$$\frac{1}{2}[(f + g)^2 - f^2 - g^2] = fg$$

is also integrable. $\qquad\square$

## 11.5   The Fundamental Theorem of the Calculus

So far we have developed some powerful theoretical tools to show that a given function is integrable and to relate the integrals of related integrable functions, but we don't have any really convenient techniques for actually computing Riemann integrals. To compute

$$\int_0^1 x^2 \, dx,$$

for example, we had to resort to exhibiting a sequence of dissections whose upper and lower sums converge to a common limit, in this case, $\frac{1}{3}$ (see Example 11.11). Theorem 11.12 tells us that, in principle, we can always compute Riemann integrals like this, but in practice this method of computing integrals is very onerous and, unless the integrand (the function to be integrated) is fairly simple, is likely to be intractable. Consider, for example, attempting to compute

$$\int_0^\pi \sin x \, dx$$

using this method. We know that this integral exists (Theorem 11.16) because we know that sin is continuous, but we have no hope of computing it directly using Riemann sums. In this section, we establish a fundamental connexion between Riemann integration and differentiation which will give us a convenient means of computing $\int_a^b f$ whenever we can dream up a function whose *derivative* equals $f$. Once we have done this we will rarely have to resort to computing sequences of Riemann sums to compute integrals.

**Theorem 11.25 (Fundamental Theorem of the Calculus version 1).**
*Let $f : [a, b] \to \mathbb{R}$ be continuous and define $F : [a, b] \to \mathbb{R}$ by*

$$F(x) = \int_a^x f.$$

*Then $F$ is differentiable, and $F' = f$.*

**Proof.** First note that $f$ is continuous on $[a, x]$ for all $x \in [a, b]$, so $F$ is well-defined by Theorem 11.16. We wish to compute

$$\lim_{y \to x} \frac{F(y) - F(x)}{y - x}.$$

So, let $(y_n)$ be any sequence in $[a, b] \backslash \{x\}$ such that $y_n \to x$, and

$$s_n = \frac{F(y_n) - F(x)}{y_n - x}.$$

We must show that $s_n \to f(x)$.

For each $n$, either $y_n > x$ or $y_n < x$. If $y_n > x$ then, by Theorem 11.17,

$$s_n = \frac{1}{y_n - x} \int_x^{y_n} f,$$

whereas if $y_n < x$,

$$s_n = \frac{1}{x - y_n} \int_{y_n}^x f.$$

In either case, by the Extreme Value Theorem, there exist $w_n$ and $z_n$ between $x$ and $y_n$ such that $f(w_n)$ is the minimum value of $f$ on the closed interval with endpoints $x$ and $y_n$, and $f(z_n)$ is the maximum value of $f$ on this interval. Hence, by Proposition 11.4,

$$\frac{1}{y_n - x} f(w_n)(y_n - x) \le s_n \le \frac{1}{y_n - x} f(z_n)(y_n - x) \qquad \text{if } y_n > x,$$

and $\quad \dfrac{1}{x - y_n} f(w_n)(x - y_n) \le s_n \le \dfrac{1}{x - y_n} f(z_n)(x - y_n) \qquad \text{if } y_n < x.$

In either case, we conclude that

$$f(w_n) \le s_n \le f(z_n).$$

Now, $y_n \to x$ so $w_n \to x$ and $z_n \to x$ also, by the Squeeze Rule, and $f$ is continuous, so $f(w_n) \to f(x)$ and $f(z_n) \to f(x)$. Hence, by the Squeeze Rule again, $s_n \to f(x)$, which completes the proof. $\qquad \square$

As a by-product of this theorem we see that any function which is continuous on a closed bounded interval is the derivative of some differentiable function on that interval, an interesting and far from obvious fact. Even better, Theorem 11.25 immediately implies a second theorem which renders the job of computing many Riemann integrals almost trivial:

**Theorem 11.26 (Fundamental Theorem of the Calculus version 2).** *Let $f : [a, b] \to \mathbb{R}$ be continuous and $F : [a, b] \to \mathbb{R}$ be any differentiable function such that $F' = f$. Then*

$$\int_a^b f = F(b) - F(a).$$

**Proof.** Define the function $g : [a, b] \to \mathbb{R}$,

$$g(x) = \left( \int_a^x f \right) - F(x).$$

Then, by Theorem 11.25 and the definition of $F$,

$$g'(x) = f(x) - f(x) = 0$$

for all $x \in [a, b]$. Hence, by Corollary 9.21, $g$ is constant, so $g(b) = g(a)$, that is

$$\left( \int_a^b f \right) - F(b) = \left( \int_a^a f \right) - F(a) = 0 - F(a),$$

and the result immediately follows. □

**Example 11.27.** Compute the Riemann integrals

(i) $\displaystyle\int_{-1}^1 x^2 \, dx,$

(ii) $\displaystyle\int_0^\pi \sin x \, dx,$

(iii) $\displaystyle\int_{-\pi}^\pi \sin x \sin(x^2) \, dx.$

*Solution:*

(i) Let $f(x) = x^2$ and $F(x) = x^3/3$. Then $f$ is continuous and $F' = f$ on $[-1, 1]$ so, by Theorem 11.26,

$$\int_{-1}^1 f = F(1) - F(-1) = \frac{2}{3}.$$

(ii) Let $f(x) = \sin x$ and $F(x) = -\cos x$. Then $f$ is continuous and $F' = f$ on $[0, \pi]$ so, by Theorem 11.26,

$$\int_0^\pi f = F(\pi) - F(0) = -\cos \pi + \cos 0 = 2.$$

(iii) Let $f(x) = \sin x \sin(x^2)$. Then $f$ is continuous on $[-\pi, \pi]$ so, by Theorem 11.25, there certainly exists a differentiable function $F : [-\pi, \pi] \to \mathbb{R}$ such that $F'(x) = f(x)$. But what is it? The trouble is, we have no idea how to write it down explicitly, and without an explicit formula for $F$, Theorem 11.26 isn't much use. To handle this integral, we need a bit more theory. □

**Corollary 11.28.** *Let $f : [a, b] \to \mathbb{R}$ be continuous and define $F : [a, b] \to \mathbb{R}$ by*

$$F(x) = \int_x^b f.$$

*Then $F$ is differentiable, and $F' = -f$.*

**Proof.** Let $G(x) = \int_a^x f$. Then, by Theorem 11.17, $G(x) + F(x) = \int_a^b f$, a constant. Hence $F'(x) = -G'(x) = -f(x)$, by Theorem 11.25. $\qquad\square$

**Proposition 11.29.** *Let $f : [-a, a] \to \mathbb{R}$ be a continuous odd function (meaning $f(-x) = -f(x)$ for all $x \in [-a, a]$). Then*

$$\int_{-a}^a f = 0.$$

**Proof.** Define $F : [0, a] \to \mathbb{R}$ by

$$F(x) = \int_{-x}^x f = \int_{-x}^0 f + \int_0^x f,$$

by Theorem 11.17. Then, by Theorem 11.25, Corollary 11.28, and the Chain Rule

$$F'(x) = -(-f(-x)) + f(x) = f(-x) + f(x) = 0,$$

since $f$ is odd. Hence, by Corollary 9.21, $F$ is contant, so

$$\int_{-a}^a f = F(a) = F(0) = 0.$$

$\qquad\square$

In fact, the assumption that $f$ is continuous in Proposition 11.29 is not necessary: $\int_{-a}^a f = 0$ for any odd, bounded, Riemann integrable function on $[-a, a]$. Proposition 11.29 suffices to finish off Example 11.27, however:

**Example 11.27(iii) (revisited).** We note that $f(x) = \sin x \sin(x^2)$ is an odd continuous function, so

$$\int_{-\pi}^\pi f = 0$$

by Proposition 11.29. $\qquad\square$

Having established Theorem 11.26, we see that we can explicitly compute

$$\int_a^b f(x) \, dx$$

if we can think of an **antiderivative** of $f$, that is, any function whose *derivative* is $f(x)$. This trick is so pervasive in integral calculus that it leads the unwary to identify *integration* of $f$ (that is, the process of computing a

Riemann integral $\int_a^b f$) with the process of writing down an antiderivative of $f$. Indeed, it is common practice to call an antiderivative of $f$ an *indefinite integral* of $f$ and to denote any such function by the symbol

$$\int f(x)\, dx.$$

This notation has some practical advantages, but it also, unfortunately, generates a huge amount of confusion. Note that

$$\int_a^b f(x)\, dx$$

is a single number and that its *definition* has nothing whatsoever to do with antiderivatives of $f$: it is the unique real number which is no greater than any upper Riemann sum of $f$ on $[a, b]$ and no less than any lower Riemann sum of $f$ on $[a, b]$. In particular, it is *not* a function of the "variable" $x$. Indeed, we could equally well have written it $\int_a^b f(y)\, dy$ or $\int_a^b f(\Gamma_{\widehat{\aleph}})\, d\Gamma_{\widehat{\aleph}}$, which is one reason to prefer the simpler notation $\int_a^b f$. It *is* a function of $a$ and $b$, however.

By contrast, $\int f(x)\, dx$ is just a (slightly ambiguous) symbol denoting any function whose derivative (at $x$) is $f(x)$. It *is* a function of $x$, not a single number, and its *definition* has nothing to do with Riemann sums. That these two things turn out to be closely related is (as its name suggests) a very important theorem: the Fundamental Theorem of the Calculus. To understand calculus properly it is important to maintain a clear conceptual distinction between the Riemann integral $\int_a^b f$ and any antiderivative that one might use to compute it.

We have now rigorously developed all the fundamental concepts of differential and integral calculus. It would be a straightforward matter to combine the Fundamental Theorem of the Calculus with the Product Rule of differentiation to develop the technique of integration by parts, or with the Chain Rule to develop the technique of integration by substitution. Such development requires no new analytic insight and so is better placed in a course or text on the methods of calculus, rather than a text on real analysis such as this.

## 11.6   Summary

- A **dissection** of $[a, b]$ is a finite subset $\mathscr{D} = \{a_0, a_1, \ldots, a_n\}$ of $[a, b]$ such that
$$a = a_0 < a_1 < a_2 < \cdots < a_n = b.$$
- Given a bounded function $f : [a, b] \to \mathbb{R}$, and a dissection $\mathscr{D}$, the **lower Riemann sum** is
$$l_{\mathscr{D}}(f) = \sum_{j=1}^{n} \inf\{f(x) \,:\, a_{j-1} \leq x \leq a_j\}(a_j - a_{j-1}),$$
and the **upper Riemann sum** is
$$u_{\mathscr{D}}(f) = \sum_{j=1}^{n} \sup\{f(x) \,:\, a_{j-1} \leq x \leq a_j\}(a_j - a_{j-1}).$$
- The **lower Riemann integral** of $f$ is
$$l(f) = \sup\{l_{\mathscr{D}}(f) \,:\, \mathscr{D} \text{ any dissection of } [a, b]\},$$
and the **upper Riemann integral** of $f$ is
$$u(f) = \inf\{u_{\mathscr{D}}(f) \,:\, \mathscr{D} \text{ any dissection of } [a, b]\}.$$
- $f$ is **Riemann integrable** if $l(f) = u(f)$, and in this case we denote their common value by
$$\int_a^b f$$
and call it the **Riemann integral** of $f$ (on, or over, $[a, b]$).
- A bounded function $f : [a, b] \to \mathbb{R}$ is Riemann integrable if and only if there exists some sequence $(\mathscr{D}_n)$ of dissections of $[a, b]$ such that $l_{\mathscr{D}_n}(f)$ and $u_{\mathscr{D}_n}(f)$ converge to a common limit (and, in this case, $\int_a^b f$ is the value of this common limit).
- We used this theorem to prove that all **continuous** functions, and all **monotonic** functions are Riemann integrable. We also used it to prove that
$$\int_a^b f + \int_b^c f = \int_a^c f, \quad \text{and} \quad \int_a^b (\alpha f + \beta g) = \alpha \int_a^b f + \beta \int_a^b g$$
where $\alpha, \beta$ are constants.
- There is an important link between Riemann integrals and derivatives, given by the **Fundamental Theorem of the Calculus**:
  - Version 1: if $f$ is continuous and $F(x) = \int_a^x f$, then $F' = f$.
  - Version 2: if $f$ is continuous and $F$ is some antiderivative of $f$ (that is, a function satisfying $F' = f$), then $\int_a^b f = F(b) - F(a)$.
- Version 2 of the Fundamental Theorem of the Calculus provides a flexible and convenient method for computing many Riemann integrals.

## 11.7 Tutorial problems

1. (a) Prove that $\sum_{j=1}^{n} j^3 = \frac{1}{4}n^2(n+1)^2$.

   (b) Use Theorem 11.12 to evaluate $\int_0^1 x^3 \, dx$ directly (without using the Fundamental Theorem of the Calculus).

2. Write down (or draw the graph of) a function $f : [a,b] \to \mathbb{R}$ and a pair of dissections $\mathscr{D}, \mathscr{D}'$ of $[a,b]$ such that $\mathscr{D}'$ is a refinement of $\mathscr{D}$, $u_{\mathscr{D}'}(f) < u_{\mathscr{D}}(f)$, but $l_{\mathscr{D}'}(f) \not> l_{\mathscr{D}}(f)$.

3. Write down a function $f : [0,1] \to \mathbb{R}$ which is *not* Riemann integrable but whose square, $f^2 : [0,1] \to \mathbb{R}$, *is*. Rigorously justify your answer.

4. Let $f : [a,b] \to \mathbb{R}$ be a Riemann integrable function and $g : [a,b] \to \mathbb{R}$ be a function which coincides with $f$, except at a finite set of points $c_1, \ldots, c_p \in [a,b]$. Prove that $g$ is Riemann integrable and that

$$\int_a^b g = \int_a^b f.$$

## 11.8 Homework problems

1. (a) Prove that $\sum_{j=1}^{n} j = \frac{1}{2}n(n+1)$.

   (b) Use Theorem 11.12 to evaluate $\int_0^1 x \, dx$ directly (without using the Fundamental Theorem of the Calculus).

2. Let $f : [0,2] \to \mathbb{R}$,

$$f(x) = \begin{cases} 1 \text{ if } x \neq 1 \\ 2 \text{ if } x = 1 \end{cases}.$$

   Use Theorem 11.12 to prove that $f$ is Riemann integrable and compute $\int_0^2 f$.

3. Let $f : [-a,a] \to \mathbb{R}$ be bounded, Riemann integrable, and odd. Prove that $\int_{-a}^a f = 0$.

4. Show that $f : \mathbb{R} \to \mathbb{R}$, $f(x) = \exp(-x^2)$ is Riemann integrable on every closed bounded interval $[a,b]$. Show that

$$\lim_{n \to \infty} \int_0^n f$$

exists and is less than $e/(e-1)$.

# Chapter 12

# Logarithms and irrational powers

## 12.1 Logarithms

In section 6.3 we showed that every postive real number $a$ has a unique $p^{\text{th}}$ root, $a^{1/p}$, and this allowed us to define $a^r$ for any rational power $r = n/p$, namely $a^{n/p} = (a^{1/p})^n$. How can we make sense of $a^x$ when $x$ is some real, possibly *irrational* number? We could take a rational sequence $r_n \to x$ (which must exist by Theorem 1.25) and define

$$a^x = \lim_{n \to \infty} a^{r_n},$$

but the technicalities involved in checking that this is well-defined are quite formidable. (How do we know the sequence above converges? If it does, how do we know its limit is independent of the choice of sequence $(r_n)$?) Luckily there is another way to define $a^x$ which avoids these technicalities, using **logarithms**. In this short chapter we define logarithms using the Riemann integral, use them to formulate a definition of $a^x$ which works for all $x \in \mathbb{R}$, and show that this new definition of $a^x$ coincides with the old one in the case where $x$ is rational.

**Definition 12.1.** The **(natural) logarithm** is the function

$$\log : (0, \infty) \to \mathbb{R}, \qquad \log x = \int_1^x \frac{1}{t} dt.$$

## Remarks

- The function $f(t) = 1/t$ is continuous on $(0, \infty)$, and hence Riemann integrable on $[1, x]$ (if $x \geq 1$) or $[x, 1]$ (if $0 < x < 1$), so the function log is well-defined (Theorem 11.16).

- log is differentiable, by the Fundamental Theorem of the Calculus, and

$$\log'(x) = \frac{1}{x}.$$

- It follows immediately from the definition that

$$\log 1 = \int_1^1 \frac{1}{t} dt = 0.$$

The logarithm function obeys a very useful identity.

**Proposition 12.2.** *For all* $x, y \in (0, \infty)$, $\log(xy) = \log x + \log y$.

**Proof.** Choose and fix $y \in (0, \infty)$ and consider the function

$$f : (0, \infty) \to \mathbb{R}, \qquad f(x) = \log(xy) - \log x - \log y.$$

By the Chain Rule,

$$f'(x) = \frac{y}{xy} - \frac{1}{x} = 0$$

so $f$ is constant (Corollary 9.21). Hence, for all $x$, $f(x) = f(1) = \log y - 0 - \log y = 0$. $\square$

A second useful identity quickly follows from this.

**Proposition 12.3.** *For all* $x \in (0, \infty)$ *and* $n \in \mathbb{Z}$, $\log x^n = n \log x$.

**Proof.** Let $x > 0$. Then $\log x^0 = \log 1 = 0$, so the claim holds for $n = 0$. Assume it holds for $n = k$, some non-negative integer. Then

$$\log x^{k+1} = \log(xx^k) = \log x + \log x^k = (k+1) \log x$$

by Proposition 12.2, so the claim also holds for $n = k + 1$. Hence, by induction, the claim holds for all $n \in \mathbb{Z}^+$.

Let $n \in \mathbb{Z}$, $n < 0$. Then

$$\log x^n = \log \left( \frac{1}{x^{|n|}} \right) = |n| \log \left( \frac{1}{x} \right),$$

as we have just proved. By Proposition 12.2,

$$0 = \log 1 = \log \left( x \times \frac{1}{x} \right) = \log x + \log \frac{1}{x},$$

so $\log(1/x) = -\log x$. Hence

$$\log x^n = -|n| \log x = n \log x,$$

as was to be shown. $\square$

**Proposition 12.4.** *The function* $\log : (0, \infty) \to \mathbb{R}$ *is smooth, strictly increasing, and bijective.*

**Proof.** Let $f : (0, \infty) \to \mathbb{R}$, $f(x) = 1/x$. We have noted that log is differentiable with derivative $f$. But $f$ is rational, and hence smooth, so log is smooth. Also, for all $x \in (0, \infty)$, $f(x) > 0$, so log is strictly increasing, and hence injective (Proposition 9.24). It remains to show that log is surjective.

For each $n \in \mathbb{Z}^+$, $n \geq 2$, let $\mathscr{D}_n$ be the regular dissection of $[1, n]$ of size $(n - 1)$, that is

$$\mathscr{D}_n = \{1, 2, \ldots, n\}.$$

Then

$$\log n = \int_1^n f \geq l_{\mathscr{D}_n}(f) = \sum_{j=1}^{n-1} (1) f(j+1) = \sum_{k=2}^n \frac{1}{k} =: s_n,$$

where we have used the fact that $f$ is monotonically decreasing. The sequence $(s_n)$ is unbounded above (Example 5.2), so the sequence $\log n$ is also unbounded above. Hence, given any $K > 0$, there exists $n \in \mathbb{Z}^+$ such that $\log n > K$. But $\log 1 = 0$, and log is continuous (since it is differentiable) so, by the Intermediate Value Theorem, there exists $x \in [1, n]$ such that $\log x = K$. Hence, log takes all non-negative values. Let $L < 0$. As we just showed, there exists $x \in [1, \infty)$ such that $\log x = -L$. But then, by Proposition 12.3, $\log(1/x) = -\log x = L$. So log also takes all negative values, and we conclude that $\log : (0, \infty) \to \mathbb{R}$ is surjective, hence bijective. $\qquad\square$

We previously used the term *natural logarithm* to mean the inverse function to the exponential function (see section 10.3). Definition 12.1 has no obvious connexion with the exponential function (which was, recall, defined using a convergent power series), so this coincidence of terminology needs to be justified. In fact log, defined as in Definition 12.1 really is the inverse function to exp:

**Proposition 12.5.** $\log : (0, \infty) \to \mathbb{R}$ *is the inverse function to* $\exp : \mathbb{R} \to (0, \infty)$.

**Proof.** We must show that, for all $x \in \mathbb{R}$, $\log(\exp(x)) = x$, and for all $y \in (0, \infty)$, $\exp(\log y) = y$. Consider the function

$$f : \mathbb{R} \to \mathbb{R}, \qquad f(x) = \log(\exp(x)) - x.$$

By the Chain Rule,

$$f'(x) = \frac{\exp(x)}{\exp(x)} - 1 = 0$$

for all $x$, so $f$ is constant (Corollary 9.21). Hence, for all $x$, $f(x) = f(0) = \log 1 - 0 = 0$, that is, $\log(\exp(x)) = x$. Now, for all $y \in (0, \infty)$,

$$\log(\exp(\log y)) = \log y,$$

as we have just shown. But log is injective, so if $\log z = \log y$ then $z = y$. Hence $\exp(\log y) = y$, as was to be shown. □

So, for each positive number $y$, $\log y$ is the real number whose exponential is $y$. Recall we defined *Euler's number* to be

$$e = \exp(1) = \sum_{n=0}^{\infty} \frac{1}{n!}.$$

It follows that $\log e = 1$. There is an alternative way to define $e$, as the limit of the sequence $(1 + \frac{1}{n})^n$. We are now in a position to prove this.

**Proposition 12.6.** *The sequence* $x_n = \left(1 + \dfrac{1}{n}\right)^n$ *converges to e.*

**Proof.** It is clear that $x_n > 0$ for all $n$. Now, for all $n$, by Proposition 12.3,

$$\log x_n = n \log \left(1 + \frac{1}{n}\right)$$
$$= \frac{\log \left(1 + \frac{1}{n}\right) - \log 1}{\left(1 + \frac{1}{n}\right) - 1}$$
$$= \frac{\log y_n - \log 1}{y_n - 1}$$

where $y_n = 1 + \frac{1}{n}$. Note that $y_n \to 1$. Since log is differentiable (everywhere, and, in particular, at 1), the sequence of difference quotients

$$\frac{\log y_n - \log 1}{y_n - 1}$$

converges to $\log'(1) = 1/1 = 1$, that is,

$$\log x_n \to 1.$$

Now exp is continuous, so

$$\exp(\log x_n) \to \exp(1) = e.$$

But $\exp(\log x_n) = x_n$ (Proposition 12.5), so $x_n \to e$. □

## 12.2 Irrational (and rational) powers

Proposition 12.5 implies that, for all $a > 0$, $\log a$ is the unique real number $y$ such that $\exp(y) = a$. This motivates the following definition:

**Definition 12.7.** Let $a > 0$ and $x \in \mathbb{R}$. Then $a$ **to the power** $x$ is
$$[a]^x = \exp(x \log a).$$

We temporarily denote this number $[a]^x$ rather than $a^x$ to distinguish it from our "old" definition of $a^x$ when $x$ is integer or rational (so $a^3 = a \times a \times a$ and $a^{1/2} = \sqrt{a}$, whereas $[a]^3 = \exp(3 \log a)$ and $[a]^{1/2} = \exp(\frac{1}{2} \log a)$). We will shortly show that $[a]^x = a^x$ when $x$ is rational, so we can use $a^x$ to denote both.

Recall (Lemma 10.13) that, for all $x, y \in \mathbb{R}$,
$$\exp(x + y) = \exp(x) \exp(y).$$

It follows almost immediately that the number $[a]^x$ respects the usual rules for exponents.

**Proposition 12.8.** *For all* $a, b \in (0, \infty)$ *and* $x, y \in \mathbb{R}$,

(i) $[a]^{x+y} = [a]^x [a]^y$,
(ii) $[a]^x [b]^x = [ab]^x$,
(iii) $[[a]^x]^y = [a]^{xy}$.

**Proof.**

(i) $\qquad [a]^{x+y} = \exp(x \log a + y \log a)$
$$= \exp(x \log a) \exp(y \log a) = [a]^x [a]^y$$

(ii) $\qquad [a]^x [b]^x = \exp(x \log a) \exp(x \log b)$
$$= \exp(x(\log a + \log b)) = \exp(x \log ab) = [ab]^x$$

(iii) $\qquad [[a]^x]^y = \exp(y \log([a]^x)) = \exp(y \log(\exp(x \log a)))$
$$= \exp(yx \log a) = [a]^{xy}.$$

$\qquad\qquad\qquad\qquad\qquad\qquad\qquad\qquad\qquad\qquad\qquad\qquad\qquad\qquad \Box$

**Proposition 12.9.** *For all* $n \in \mathbb{Z}$ *and* $x \in \mathbb{R}$,
$$\exp(nx) = (\exp(x))^n.$$

**Proof.** Let $y = \exp(x)$. Then $x = \log y$, so
$$\exp(nx) = \exp(n \log y) = \exp(\log(y^n)) = y^n = (\exp(x))^n$$
by Proposition 12.3.

$\qquad\qquad\qquad\qquad\qquad\qquad\qquad\qquad\qquad\qquad\qquad\qquad\qquad\qquad \Box$

We can now show that, for rational $x$, $[a]^x = a^x$:

**Proposition 12.10.** *Let $p \in \mathbb{Z}$, $q \in \mathbb{Z}^+$, and $a > 0$. Then*

$$[a]^{p/q} = a^{p/q}.$$

**Proof.** Recall that $a^{p/q}$ is, by definition, $(a^{1/q})^p$. By Proposition 12.8,

$$[a]^{p/q} = [a]^{\frac{1}{q} \times p} = [[a]^{1/q}]^p.$$

Now

$$([a]^{1/q})^q = (\exp(\frac{1}{q} \log a))^q = \exp(\log a) = a$$

by Propositions 12.9 and 12.5, so $[a]^{1/q}$ is a positive real number whose $q^{\text{th}}$ power is $a$. But this number is unique and is, by definition, $a^{1/q}$. Hence

$$[a]^{p/q} = [a^{1/q}]^p = \exp(p \log(a^{1/q})) = (\exp(\log(a^{1/q})))^p = (a^{1/q})^p$$

by Propositions 12.9 and 12.5. $\square$

From now on we cease to distinguish between $[a]^x$ and $a^x$. Recall Proposition 9.15, which said that the function $f(x) = x^r$, where $r$ is a rational constant, is differentiable on $(0, \infty)$ and has derivative $f'(x) = rx^{r-1}$. We are now in a position to prove something more general: that the same assertion holds for $r$ any *real* constant.

**Proposition 12.11.** *Let $f : (0, \infty) \to \mathbb{R}$ be the function $f(x) = x^r$, where $r$ is any real constant. Then $f$ is differentiable, and for all $x \in (0, \infty)$,*

$$f'(x) = rx^{r-1}.$$

**Proof.** We note that $f(x) = \exp(r \log x)$ and so, by the Chain Rule, $f$ is differentiable with

$$f'(x) = r \exp'(r \log x) \log'(x) = r \exp(r \log x) \frac{1}{x} = r \frac{x^r}{x} = rx^{r-1}.$$

$\square$

## 12.3 Summary

- The **natural logarithm** function is

$$\log : (0, \infty) \to \mathbb{R}, \qquad \log x = \int_1^x \frac{1}{t} dt.$$

- log is smooth, increasing, and bijective.
- Its inverse function is $\exp : \mathbb{R} \to (0, \infty)$.
- For all $x, y \in (0, \infty)$,

$$\log(xy) = \log x + \log y.$$

- For all $a > 0$ and $x \in \mathbb{R}$ we define $a$ **to the power** $x$ to be

$$a^x = \exp(x \log a).$$

This satisfies all the usual algebraic rules of exponents and coincides with $(a^{1/q})^p$ when $x = p/q$ is rational.

## 12.4   Tutorial problems

1. Let $b > 1$ be a constant, and $f : \mathbb{R} \to (0, \infty)$ such that $f(x) = b^x$. Show that $f$ is smooth, strictly increasing, and bijective. It follows that $f$ has an inverse function, which we denote $\log_b : (0, \infty) \to \mathbb{R}$, and call the **logarithm to the base** $b$. Show that

   (a) for all $x \in (0, \infty)$, $\log_b x = \frac{\log x}{\log b}$,

   (b) for all $x, y \in (0, \infty)$, $\log_b(xy) = \log_b x + \log_b y$,

   (c) for all $x, y \in (0, \infty)$, $\log_b(x^y) = y \log_b x$,

   (d) $\log_b$ is also smooth and strictly increasing.

   Compute $\log_b 1$, $\log_b b$, and $\log_b e$.

   *Remark: note that* $\log_e = \log$.

2. Prove that the sequence $x_n = \frac{\log n}{n}$ converges to 0. Deduce that $y_n = n^{1/n} \to 1$.

3. Prove that the series $\displaystyle\sum_{n=2}^{\infty} \frac{1}{n \log n}$ diverges.

## 12.5   Homework problems

1. Let $r > 0$ be constant. Prove that the sequence $x_n = \frac{\log n}{n^r}$ converges to 0.

   *Remark: this is true no matter how small the positive constant r is. We deduce the useful fact that "power law growth always beats logarithmic growth".*

2. Prove that the series $\displaystyle\sum_{n=2}^{\infty} \frac{\log n}{n^2}$ converges.

3. Let $r > 0$ be constant. Prove that the series $\displaystyle\sum_{n=2}^{\infty} \frac{1}{n^r}$ converges if and only if $r > 1$.

# Chapter 13

# What are the reals?

So far, we have been rather coy about what the real numbers actually are. We just declared that $\mathbb{R}$, the set of reals, is a complete ordered field, that is, a set on which the basic operations of arithmetic are well-defined (a *field*), which is equipped with a consistent means of determining which of two different numbers is the larger (an *ordering relation*) and satisfies Axiom 1.22 (the *Axiom of Completeness*). We were able to show that any such field is uncountable, contains a copy of the set of rationals $\mathbb{Q}$ as a dense subset, and has the Archimedean Property (it contains a copy of $\mathbb{Z}^+$ as an unbounded subset). There remains, however, a worrying conceptual gulf between our axiomatic treatment of $\mathbb{R}$ and the more informal ways we usually think about real numbers – as points on a number line, or (possibly infinite) decimal strings, for example. Even worse, we haven't actually shown that $\mathbb{R}$ *exists*! That is, we have shown that *if* a complete ordered field exists then it must have a long list of interesting properties, and we have used these to develop a theory of convergence of sequences, continuity of functions, and differential and integral calculus in this field. But what if no such field exists? Then the whole exercise becomes vacuous! The purpose of this final chapter is to fill in this yawning gap.

There are at least two ways to give a constructive definition of $\mathbb{R}$, both of which start with $\mathbb{Q}$. One very elegant definition takes a real number to be a partition of $\mathbb{Q}$ into a pair of disjoint subsets satisfying certain properties. Such a partition is called a *Dedekind cut*. The philosophy of this book is to formulate all fundamental notions in terms of *sequences*, however, so it is more natural for us to follow the alternative route, and define real numbers as (equivalence classes of) rational sequences. This has several advantages besides philosophical coherence. The general strategy can be adapted to other settings, to produce other interesting completions of $\mathbb{Q}$, and completions of other, more elaborate objects in functional analysis. It allows us to introduce a fundamental convergence-like property of sequences (the Cauchy property) which is of independent interest. It also allows us to make direct contact with the informal notion of real numbers as decimal expansions. For the purposes of motivation, this is where we begin.

My pocket calculator tells me that

$$\sqrt{2} = 1.4142135623730950488\cdots.$$

What does this mean? We know that, if $\mathbb{R}$ exists, it contains a unique positive element whose square is 2, and my calculator is asserting that this number is the limit of a sequence of rational numbers beginning

$$1, \ 1.4, \ 1.41, \ 1.414, \ 1.4142, \ 1.41421, \ldots.$$

It's tempting to define $\sqrt{2}$ to *be* this sequence of rational numbers. But there's a problem. Although my calculator speaks to me in decimal, because it knows I'm human and probably have 10 fingers, it really thinks that

$$\sqrt{2} = 1.0110101000001001111\cdots$$

because it actually thinks in binary. So what it really knows is that $\sqrt{2}$ is the limit of a completely different rational sequence beginning

$$1, \ 1, \ 1 + \frac{1}{4}, \ 1 + \frac{1}{4} + \frac{1}{8}, \ 1 + \frac{1}{4} + \frac{1}{8}, \ 1 + \frac{1}{4} + \frac{1}{8} + \frac{1}{32}, \ldots.$$

In fact, there are infinitely many different rational sequences which converge to $\sqrt{2}$ (see Homework Problem 2 in Chapter 3 for another example, defined inductively). What we need is some way of defining when two rational sequences are equivalent, in that they define the same real number. We can then *define* a real number to be a set of equivalent rational sequences (so $\sqrt{2}$ would be an infinite set of rational sequences which includes the two particular sequences above). But we can't just say "two rational sequences are equivalent if they converge to the same limit", because our definition of convergence already presupposes the existence of a limit in $\mathbb{R}$, so our definition of real number would be circular (it would contain reference to the real numbers themselves). We can, however, say that two rational sequences $(x_n), (y_n)$ are equivalent if $x_n - y_n$ converges to 0, since convergence to the rational number 0 can be defined without any reference to $\mathbb{R}$ (as we will see). So perhaps we can define a real number to be any set of equivalent rational sequences? The problem with this is that *every* rational sequence would then define a real number, even something crazy like

$$x_n = (-1)^n n$$

which can't really be thought of as converging to anything sensible. What we need, then, is a notion of convergence for rational sequences which makes no reference to the idea of *limit*. This is provided by the *Cauchy property*.

But before we launch into defining the reals, perhaps we should answer a more basic question. What, in fact, are the *rationals*?

## 13.1  What are the rationals?

At elementary school we are taught to think of $\frac{2}{5}$ as meaning 2 lots of 5 equal portions of a single object (usually, but not always, a pie). If we are to do away with the platonic notion of perfectly symmetric pies, cut into perfectly identical portions, how should we define $\frac{2}{5}$? An alternative is to say that, if one had 5 pies then $\frac{2}{5}$ would be two of them, if one had 10, then $\frac{2}{5}$ would be 4 of them, if one had 15, $\frac{2}{5}$ would be 6 etc. This avoids the conceptually tricky notion of dividing 1 into smaller pieces, at the cost of identifying $\frac{2}{5}$ with an infinite set of pairs of integers: $(2,5)$, $(4,10)$, $(6,15)$.... To make this precise (and economical) we need to introduce the idea of an *equivalence relation* on a set, an idea which completely pervades pure mathematics.

**Definition 13.1.** A **relation** on a set $A$ is a function $R : A \times A \to \{0,1\}$. That is, it is a rule which assigns to each ordered pair $(x,y)$, where $x, y \in A$, one of the two values 0 or 1. We think of 0 as representing "false" and 1 as representing "true". If $R(x,y) = 1$, we say that "$x$ is related to $y$ under $R$" and denote this $xRy$. If $R(x,y) = 0$, we say that "$x$ is not related to $y$ under $R$" and denote this $x \not\mathrel{R} y$.

**Example 13.2.** $<$ is a relation on $\mathbb{Z}$. It maps $(1,2)$ to 1 and $(3,3)$ to 0. That is $1 < 2$, but $3 \not< 3$.

**Definition 13.3.** A relation $R$ on $A$ is an **ordering relation** if it has the following properties:

(i) Irreflexivity: for all $x \in A$, $x \not\mathrel{R} x$.
(ii) Transitivity: for all $x, y, z \in A$, if $xRy$ and $yRx$ then $xRz$.
(iii) Trichotomy: for all $x, y \in A$ exactly one of the following holds: $xRy$, $x = y$, or $yRx$.

The relation $<$ on $\mathbb{Z}$ is an ordering relation. Of more immediate interest is a different kind of relation:

**Definition 13.4.** A relation $R$ on $A$ is an **equivalence relation** if it has the following properties:

(i) Reflexivity: for all $x \in A$, $xRx$.
(ii) Transitivity: for all $x, y, z \in A$, if $xRy$ and $yRx$ then $xRz$.
(iii) Symmetry: for all $x, y \in A$, if $xRy$ then $yRx$.

Clearly $<$ on $\mathbb{Z}$ is *not* an equivalence relation. For example, $1 \not< 1$ so it isn't reflexive, and $1 < 2$ but $2 \not< 1$, so it isn't symmetric.

**Example 13.5.** Define a relation $R$ on $\mathbb{Z}$ by the condition that $R(n, m) = 1$ if and only if $n - m$ is divisible by 2. This is an equivalence relation (check it!).

We usually denote equivalence relations by the symbol $\sim$ rather than $R$, and if $x \sim y$, we say that "$x$ is equivalent to $y$ (under $\sim$)". So, in Example 13.5, 1 is equivalent to $-1$ and 5 and 1017, while 0 is equivalent to $-802$ and 46.

**Definition 13.6.** Let $\sim$ be an equivalence relation on $A$ and $x \in A$. The **equivalence class** of $x$ is the subset

$$[x] = \{y \in A \ : \ x \sim y\} \subseteq A.$$

The **quotient of $A$ by** $\sim$ is the set consisting of all equivalence classes

$$A/\sim = \{[x] \ : \ x \in A\}.$$

In Example 13.5, there are precisely two equivalence classes, namely the class of 0

$$[0] = \{\ldots, -4, -2, 0, 2, 4 \ldots\}$$

consisting of all even integers, and the class of 1

$$[1] = \{\cdots, -3, -1, 1, 3, \ldots\}$$

consisting of all odd integers. Hence $\mathbb{Z}/\sim = \{[0], [1]\}$, a set containing exactly two objects, $[0]$ and $[1]$. Note that these two classes are disjoint, $[0] \cap [1] = \emptyset$ and cover $\mathbb{Z}$, $[0] \cup [1] = \mathbb{Z}$. As we will see, this is no accident. Note also that it was an entirely arbitrary choice to call them the classes of 0 and 1: one could equally well label each of them by any other of their elements, so $[0] = [2] = [-26]$, and $[1] = [17] = [-5]$.

**Proposition 13.7.** *Let $\sim$ be an equivalence relation on $A$. Then the equivalence classes of $\sim$ form a partition of $A$, that is, they are disjoint, and their union is $A$.*

**Proof.** Let $x \in A$. Then $x \sim x$ (since $\sim$ is reflexive), so $x \in [x]$. Hence every $x \in A$ lies in some equivalence class (namely, its own), so the union of all equivalence classes is $A$. It remains to show that the classes are disjoint. So, assume $[x] \cap [y] \neq \emptyset$. Then there exists $z \in A$ such that $z \in [x]$ and $z \in [y]$, that is $x \sim z$ and $y \sim z$. We will show that, in this case, $[x] = [y]$. Let $x' \in [x]$. Then $x \sim x'$, and $\sim$ is symmetric, so $x' \sim x$. Also $x \sim z$ and $\sim$ is transitive, so $x' \sim z$. But $y \sim z$ so, since $\sim$ is symmetric, $z \sim y$ whence (since $\sim$ is transitive), $x' \sim y$ and $y \sim x'$ ($\sim$ is symmetric). Hence $x' \in [y]$. This is true for all $x' \in [x]$, so $[x] \subseteq [y]$. An identical argument shows that $[y] \subseteq [x]$, and hence $[x] = [y]$. So, if two equivalence classes overlap, they are equal. Equivalently, every pair of distinct equivalence classes is disjoint. $\square$

**Remark.** So, to every equivalence relation on $A$ there is an associated partition of $A$ into equivalence classes. In fact, the process works both ways: to every partition of $A$ there is an associated equivalence relation. We just say that $x \sim y$ if and only if $x$ and $y$ lie in the same component of the partition.

The main point of equivalence relations is that we can use structures on $A$ to define structures on $A/\sim$, provided the structure on $A$ respects the equivalence classes.

**Example 13.5 (revisited).** On $\mathbb{Z}$ we have the algebraic operations of addition and multiplication. We can use these to equip $\mathbb{Z}/\sim = \{[0], [1]\}$ with operations of addition and multiplication as follows:

$$[n] + [m] = [n + m], \qquad [n][m] = [nm].$$

We need to be careful here: in forming $[n+m]$ and $[nm]$ we have (completely arbitrarily) chosen representative elements $n \in [n]$ and $m \in [m]$. In order for this to be well-defined, we need to know that the *equivalence classes* $[n + m]$ and $[nm]$ are independent of these arbitrary choices. Let's check. Let $n' \in [n]$ and $m' \in [m]$. Then, by the definition of equivalence, there exist $k, p \in \mathbb{Z}$ such that

$$n - n' = 2k, \qquad m - m' = 2p.$$

Hence

$$(n + m) - (n' + m') = 2(k + p), \qquad nm - n'm' = 2(km + pn - 2kp)$$

and $k + p \in \mathbb{Z}$, $km + pn - 2kp \in \mathbb{Z}$, so $n' + m' \sim n + m$ and $n'm' \sim nm$. Hence $[n' + m'] = [n + m]$ and $[n'm'] = [nm]$, and all is well. $\qquad\square$

**Definition 13.8.** On the set $\mathbb{Z} \times \mathbb{Z}^+$, define the relation $\sim$ by

$$(n, m) \sim (n', m') \qquad \text{if and only if} \qquad nm' = n'm.$$

This is an equivalence relation. The set of **rational numbers** $\mathbb{Q}$ is the quotient set $(\mathbb{Z} \times \mathbb{Z}^+)/\sim$. On $\mathbb{Q}$ we define addition by

$$[(m, n)] + [(k, p)] = [(pm + kn, np)]$$

and multiplication by

$$[(m, n)] \times [(k, p)] = [(mk, np)].$$

**Exercise 13.9.**

(i) Check that $\sim$ is an equivalence relation on $\mathbb{Z} \times \mathbb{Z}^+$.
(ii) Check that the addition and multiplication operations on $\mathbb{Q}$ are well-defined (that is, independent of the choices of representatives of the equivalence classes used to define them).

It is not difficult to show that $(\mathbb{Q}, +, \times)$ is a *field*.

**Definition 13.10.** A **field** is a set $\mathbb{F}$ equipped with a pair of commutative, associative binary operations $+, \times$ such that

(i) $\times$ distributes over $+$: for all $x, y, z \in \mathbb{F}$, $x \times (y + z) = (x \times y) + (x \times z)$.
(ii) $+, \times$ have identity elements $0_\mathbb{F}, 1_\mathbb{F}$, respectively: for all $x \in \mathbb{F}$, $x + 0_\mathbb{F} = x$ and $x \times 1_\mathbb{F} = x$.
(iii) For all $x \in \mathbb{F}$ there exists $-x \in \mathbb{F}$ such that $x + (-x) = 0_\mathbb{F}$.
(iv) For all $x \in \mathbb{F} \setminus \{0_\mathbb{F}\}$, there exists $x^{-1} \in \mathbb{F}$ such that $x \times x^{-1} = 1_\mathbb{F}$.

In the case of $\mathbb{Q}$, we have

$$0_\mathbb{Q} = [(0, 1)]$$
$$1_\mathbb{Q} = [(1, 1)]$$
$$-[(m, n)] = [(-m, n)]$$
$$[(m, n)]^{-1} = \begin{cases} [(n, m)] & \text{if } m > 0, \\ [(-n, -m)] & \text{if } m < 0. \end{cases}$$

Note that, in a field, we have well-defined laws of subtraction and division, namely

$$x - y := x + (-y), \qquad x \div y := x \times y^{-1}.$$

Exercise 13.11. Show that the quotient set $\mathbb{Z}/\sim$ and addition and multiplication laws defined in Example 13.5 (revisited) form a field. Note that $\mathbb{Z}$ itself is *not* a field (since 2, for example, has no multiplicative inverse $2^{-1}$ in $\mathbb{Z}$).

Even better, $\mathbb{Q}$ is an *ordered field*:

**Definition 13.12.** An **ordered field** is a field $\mathbb{F}$ equipped with a subset $\mathbb{F}_+ \subset \mathbb{F}$, the set of **positive elements**, having the properties:

(i) For all $x, y \in \mathbb{F}_+$, $x + y \in \mathbb{F}_+$ and $x \times y \in \mathbb{F}_+$.
(ii) For all $x \in \mathbb{F}$ exactly one of the following is true: $x \in \mathbb{F}_+$, $x = 0$, or $-x \in \mathbb{F}_+$.

On the ordered field $(\mathbb{F}, \mathbb{F}_+)$, we define a relation $<_\mathbb{F}$ by

$$x <_\mathbb{F} y \qquad \text{if and only if} \qquad y - x \in \mathbb{F}_+.$$

Exercise 13.13. Show that, on any ordered field $(\mathbb{F}, \mathbb{F}_+)$, $<_\mathbb{F}$ is an ordering relation. Show further that it has the properties

(i) For all $x, y, z \in \mathbb{F}$, if $x <_\mathbb{F} y$ then $x + z <_\mathbb{F} y + z$.
(ii) For all $x, y, z \in \mathbb{F}$, if $x <_\mathbb{F} y$ and $0_\mathbb{F} <_\mathbb{F} z$ then $xz <_\mathbb{F} yz$.
(iii) For all $x, y, z \in \mathbb{F}$, if $x <_\mathbb{F} y$ and $z <_\mathbb{F} 0_\mathbb{F}$ then $yz <_\mathbb{F} xz$.

To make $\mathbb{Q}$ into an *ordered* field we define

$$\mathbb{Q}_+ = \{[(m, n)] \in \mathbb{Q} : m > 0\}.$$

Exercise 13.14. Check that $(\mathbb{Q}, \mathbb{Q}_+)$ is an ordered field (i.e. $\mathbb{Q}_+$ has the properties required by Definition 13.12).

Henceforth, we denote the class $[(m, n)] \in \mathbb{Q}$ by $\frac{m}{n}$, as usual, and the ordering relation $<_\mathbb{Q}$ by $<$. The equivalence relation defining $\mathbb{Q}$ was chosen precisely so that

$$\frac{m}{n} = \frac{m'}{n'} \qquad \text{if and only if} \qquad mn' = m'n.$$

Further, multiplication and addition were defined so that

$$\frac{m}{n} + \frac{k}{p} = \frac{mp + nk}{np}, \qquad \frac{m}{n} \times \frac{k}{p} = \frac{mk}{np}.$$

So the field $\mathbb{Q}$ has exactly the algebraic operations one is familar with.

Note that $\mathbb{Q}$ carries a copy of $\mathbb{Z}$ (with its $+$ and $\times$ operations and ordering relation $<$) embedded within it. We just identify the integer $n$ with the rational number $\frac{n}{1} = [(n, 1)]$. With this in mind, we state and prove:

**Proposition 13.15 (The Archimedean Property of $\mathbb{Q}$).** *The subset $\mathbb{Z}^+ \subset \mathbb{Q}$ is unbounded above. That is, for all $x \in \mathbb{Q}$, there exists $k \in \mathbb{Z}^+$ such that $k > x$.*

**Proof.** Let $x \in \mathbb{Q}$. Then $x = \frac{m}{n}$ for some $m \in \mathbb{Z}$ and $n \in \mathbb{Z}^+$. Let $k = |m| + 1 \in \mathbb{Z}^+$. Then

$$k - x = \frac{k}{1} - \frac{m}{n} = \frac{kn - m}{n} = \frac{|m|n + n - m}{n}.$$

Now $n \geq 1$, so $|m|n - m \geq |m| - m \geq 0$, and $n > 0$, so $|m|n + n - m > 0$. Hence $k - x \in \mathbb{Q}_+$, that is, $k > x$. $\qquad\square$

Now that we have a precise definition of $\mathbb{Q}$ (it's the quotient set $\mathbb{Z} \times \mathbb{Z}^+ / \sim$ where $\sim$ is the equivalence relation in Definition 13.8), we can turn to the main job at hand: giving a constructive definition of $\mathbb{R}$.

## 13.2   The Cauchy property

We will define real numbers to be equivalence classes of *rational sequences* under a certain equivalence relation. So, to begin with, the only numbers we are allowed to consider are rational. It is easy to adapt Definition 2.5 to give a precise meaning to the idea that a *rational* sequence converges to a *rational* number.

**Definition 13.16.** A rational sequence $(x_n)$ converges to a rational number $L$ if, for each $\varepsilon \in \mathbb{Q}_+$, there exists $N \in \mathbb{Z}^+$ such that, for all $n \geq N$, $|x_n - L| < \varepsilon$.

The only modification is that $\varepsilon$ and the limit are required to be rational. All of the limit theorems is section 3.1 extend to this (more restricted) setting, the proofs being identical (but with $\varepsilon$ and the limits required to be rational). In particular, note that Theorem 3.5, the *Algebra of Limits*, applies equally well to rational convergent sequences. The Monotone Convergence Theorem (Corollary 3.14) does *not* extend to the purely rational setting, however, because its proof makes use of the Axiom of Completeness (Axiom 1.22), which is false for $\mathbb{Q}$.

The set of convergent rational sequences is too small for our purposes: we need to include sequences which don't converge to any rational number. The correct convergence-like property, which avoids any mention of limits, was formulated by Cauchy:

**Definition 13.17.** A rational sequence $(x_n)$ is **Cauchy** (or **has the Cauchy property**) if, for each $\varepsilon \in \mathbb{Q}_+$, there exists $N \in \mathbb{Z}^+$ such that, for all $m, n \geq N$, $|x_m - x_n| < \varepsilon$. We denote the set of all rational Cauchy sequences by $\mathscr{C}$.

So, if a sequence is Cauchy, then given any positive (rational) number $\varepsilon$, there is a point in the sequence (the $N^{\text{th}}$ term) beyond which all terms lie closer than $\varepsilon$ to *one another*. This implies, in particular, that for all $n \geq N$, $|x_n - x_N| < \varepsilon$. Hence, all points on the graph of $x_n$ against $n$ to the right of the vertical line $x = N$ lie in the horizontal strip of width $2\varepsilon$ centred on $y = x_N$. It does *not* follow that $x_n \to x_N$ however, because if we change $\varepsilon$ then $N$ (in general) changes, so the line $y = x_N$ on which the narrow strip is centred moves. We *cannot* conclude that a rational Cauchy sequence converges (to any rational number) therefore, although it is straightforward to show that the converse is true: every convergent sequence is Cauchy. We leave this as an exercise. Of more immediate concern to us is the fact that the set of rational Cauchy sequences is closed under addition and multiplication.

**Lemma 13.18.** *If $(x_n)$ is Cauchy, then $(x_n)$ is bounded (above and below, in $\mathbb{Q}$).*

**Proof.** Since $(x_n)$ is Cauchy, there exists $N \in \mathbb{Z}^+$ such that, for all $n, m \geq N$, $|x_n - x_m| < 1$. Hence, for all $n \geq N$, $|x_n - x_N| < 1$, so $|x_n| < |x_N| + 1$. Consider the finite set

$$A = \{|x_1|, |x_2|, \ldots, |x_{N-1}|, |x_N| + 1\} \subset \mathbb{Q}.$$

Since it is finite, it has a maximum element, $K$ say. Then, for all $n \in \mathbb{Z}^+$, $|x_n| \leq K$. $\qquad\square$

Note that this proof is closely modelled on the proof of Proposition 3.3.

**Proposition 13.19.** *Let $(x_n), (y_n)$ be rational Cauchy sequences. Then $(x_n + y_n)$ and $(x_n y_n)$ are rational Cauchy sequences.*

**Proof.** Exercise: just modify the proof of Theorem 3.5. $\qquad\square$

Very loosely, to say that a rational sequence is Cauchy means that the terms of the sequence get arbitrarily close together if one goes sufficiently far along the sequence. In particular, it certainly follows that, for any Cauchy sequence, the sequence of differences of consecutive terms, $(x_{n+1} - x_n)$, converges to 0 (prove it!). The converse is false, however. That is, just because $x_{n+1} - x_n \to 0$, it does not follow that $(x_n)$ is Cauchy.

**Counterexample 13.20.** Consider the sequence of partial sums of the (rational) series $\sum_{n=1}^{\infty} \frac{1}{n}$, that is,

$$x_n = \sum_{k=1}^{n} \frac{1}{k}.$$

This is a rational sequence, and $x_{n+1} - x_n = \frac{1}{n+1} \to 0$. However, as we have seen (Example 5.2), $(x_n)$ is unbounded above, so cannot be Cauchy, by Lemma 13.18.                                                                    □

Just knowing that $x_{n+1} - x_n \to 0$ isn't enough to conclude that $(x_n)$ is Cauchy. If we know that $x_{n+1} - x_n \to 0$ "exponentially fast", we *can* conclude that $(x_n)$ is Cauchy, however. We next prove this in a special case, for which we will have later use.

**Lemma 13.21.** *Let $(x_n)$ be a rational sequence such that, for all $n \in \mathbb{Z}^+$,*

$$|x_{n+1} - x_n| \le \frac{1}{2^n}.$$

*Then $(x_n)$ is Cauchy.*

**Proof.** Choose and fix $\varepsilon \in \mathbb{Q}_+$. By the Archimedean Property of $\mathbb{Q}$ (Proposition 13.15), there exists $N \in \mathbb{Z}^+$ such that $N > 2/\varepsilon$, and hence $2/2^N \le 2/N < \varepsilon$. Consider all $n, m \ge N$. Without loss of generality, we may assume that $n \le m$, and hence $m = n + k$ for some $k \in \mathbb{Z}$, $k \ge 0$. Then

$$
\begin{aligned}
|x_n - x_m| &= |x_n - x_{n+1} + x_{n+1} - x_{n+2} + \cdots + x_{n+k-1} - x_{n+k}| \\
&\le |x_n - x_{n+1}| + |x_{n+1} - x_{n+2}| + \cdots + |x_{n+k-1} - x_{n+k}| \\
&\le \frac{1}{2^n} + \frac{1}{2^{n+1}} + \cdots + \frac{1}{2^{n+k-1}} \\
&= \frac{1}{2^n}\left(1 + \frac{1}{2} + \cdots + \frac{1}{2^{k-1}}\right) \\
&< \frac{2}{2^n}.
\end{aligned}
$$

Hence, since $n \ge N$, $|x_n - x_m| < 2/2^N < \varepsilon$. Hence, $(x_n)$ is Cauchy.  □

In order to equip $\mathbb{R}$ with an ordering relation, we will need a suitable definition of what it means for a Cauchy sequence to be positive.

**Definition 13.22.** Let $(x_n) \in \mathscr{C}$. Then $(x_n)$ is

(i) **null** if it converges to 0;
(ii) **strongly positive** if there exist $N \in \mathbb{Z}^+$ and $\varepsilon \in \mathbb{Q}_+$ such that, for all $n \geq N$, $x_n > \varepsilon$;
(iii) **strongly negative** if there exist $N \in \mathbb{Z}^+$ and $\varepsilon \in \mathbb{Q}_+$ such that, for all $n \geq N$, $-x_n > \varepsilon$.

We denote the set of null sequences by $\mathscr{N}$, the set of strongly positive sequences by $\mathscr{C}_+$, and the set of strongly negative sequences by $\mathscr{C}_-$. Note that $(x_n) \in \mathscr{C}_-$ if and only if $(-x_n) \in \mathscr{C}_+$.

Note that *strongly positive* is in some sense both weaker and stronger than the obvious notion of "positive Cauchy sequence", namely a Cauchy sequence $(x_n)$ such that $x_n > 0$ for all $n \in \mathbb{Z}^+$. It is weaker, since there is no reason why a finite collection of terms at the head of the sequence shouldn't be negative. For example, $(x_n) = (2n/(2n-15))$ is strongly positive (since, for all $n \geq 8$, $x_n > 1$) but it is not "positive" in the obvious sense, since $x_1, x_2, \ldots, x_7$ are all negative. On the other hand, in another way, the notion of strongly positive is stronger, since it is possible for a positive Cauchy sequence to fail to be strongly positive. For example $(x_n) = (1/n)$ is clearly positive, but is certainly not strongly positive, since it converges to 0. That a null sequence cannot be strongly positive is one consequence of the next proposition.

**Proposition 13.23.** *Let* $(x_n) \in \mathscr{C}$. *Then exactly one of the following holds:*

$$(x_n) \in \mathscr{C}_+, \qquad (x_n) \in \mathscr{N}, \qquad or \quad (x_n) \in \mathscr{C}_-.$$

**Proof.** It is clear that $(x_n)$ cannot be in both $\mathscr{C}_+$ and $\mathscr{C}_-$ (this would imply the existence $\varepsilon, \varepsilon' \in \mathbb{Q}_+$ with $-\varepsilon' > \varepsilon$, contradicting the closure of $\mathbb{Q}_+$ under addition). It is also clear that, if $(x_n) \in \mathscr{N}$, then $(x_n) \notin \mathscr{C}_+$ and $(x_n) \notin \mathscr{C}_-$. So it suffices to show that, if $(x_n) \notin \mathscr{N}$ then it is in either $\mathscr{C}_+$ or $\mathscr{C}_-$ or, equivalently, the contrapositive: if $(x_n)$ is Cauchy but in neither $\mathscr{C}_+$ nor $\mathscr{C}_-$ then it must be null.

So, assume $(x_n) \in \mathscr{C}$ but $(x_n) \notin \mathscr{C}_+$ and $(x_n) \notin \mathscr{C}_-$. Choose and fix $\varepsilon \in \mathbb{Q}_+$. Since $(x_n)$ is Cauchy, there exists $N \in \mathbb{Z}^+$ such that, for all $n, m \geq N$, $|x_n - x_m| < \frac{\varepsilon}{2}$. Since $(x_n) \notin \mathscr{C}_+$, for all $\varepsilon' \in \mathbb{Q}_+$ and all $N' \in \mathbb{Z}^+$ there exists $p \geq N'$ such that $x_p \leq \varepsilon'$. This is true, in particular, for the choice $\varepsilon' = \frac{\varepsilon}{2}$ and $N' = N$: there exists $p \geq N$ such that $x_p \leq \frac{\varepsilon}{2}$. Similarly, since $(x_n) \notin \mathscr{C}_-$, there exists $q \geq N$ such that $x_q \geq -\frac{\varepsilon}{2}$. Hence, for all $n \geq N$,

$$x_n = x_n - x_p + x_p \leq |x_n - x_p| + x_p < \frac{\varepsilon}{2} + \frac{\varepsilon}{2} = \varepsilon,$$

and

$$x_n = x_n - x_q + x_q \geq -|x_n - x_q| + x_q > -\frac{\varepsilon}{2} + \left(-\frac{\varepsilon}{2}\right) = -\varepsilon.$$

Hence, for all $n \geq N$, $|x_n| < \varepsilon$. Since $\varepsilon$ was arbitrary, it follows that $x_n \to 0$, that is, $(x_n)$ is null. $\qquad\square$

So the set of rational Cauchy sequences splits into three disjoint subsets: $\mathscr{C}_+$, $\mathscr{N}$, and $\mathscr{C}_-$. It is straightforward to show that each of these subsets is preserved under addition of null sequences:

**Proposition 13.24.** *Let $(x_n) \in \mathscr{C}$ and $(y_n) \in \mathscr{N}$.*

*(i)* *If $(x_n) \in \mathscr{C}_+$ then $(x_n + y_n) \in \mathscr{C}_+$.*
*(ii)* *If $(x_n) \in \mathscr{N}$ then $(x_n + y_n) \in \mathscr{N}$.*
*(iii)* *If $(x_n) \in \mathscr{C}_-$ then $(x_n + y_n) \in \mathscr{C}_-$.*

**Proof.** Exercise. Look to the proof of Theorem 3.5 for inspiration. $\qquad\square$

## 13.3   A sequential construction of the reals

Having developed the idea of rational Cauchy sequences, we are finally in a position to give a constructive definition of the field of real numbers. The idea is that a real number is an equivalence class of rational Cauchy sequences, where two sequences are defined to be equivalent if they differ by a null sequence.

**Definition 13.25.** On the set $\mathscr{C}$ of rational Cauchy sequences we define a relation $\sim$ as follows:

$$(x_n) \sim (y_n) \qquad \text{if and only if} \qquad (x_n - y_n) \in \mathscr{N}$$

where, as before, $\mathscr{N}$ denotes the set of rational null sequences (those which converge to 0).

**Proposition 13.26.** *The relation just defined is an equivalence relation.*

**Proof.** *Reflexivity:* for any sequence $(x_n)$, $(x_n - x_n) = (0)$ which is clearly null, so $(x_n) \sim (x_n)$.

*Symmetry:* let $(x_n) \sim (y_n)$. Then $(h_n) = (x_n - y_n)$ is null. But then $(-h_n)$ is also null (by the Algebra of Limits), so $(y_n) \sim (x_n)$.

*Transitivity:* let $(x_n) \sim (y_n)$ and $(y_n) \sim (z_n)$. Then $(h_n) = (x_n - y_n)$ and $(h'_n) = (y_n - z_n)$ are both null. But then $(h_n + h'_n) = (x_n - z_n)$ is also null (again by the Algebra of Limits), so $(x_n) \sim (z_n)$. $\qquad\square$

Since $\sim$ is an equivalence relation, it partitions $\mathscr{C}$ into disjoint equivalence classes (Proposition 13.7). We define real numbers to be precisely these disjoint equivalence classes.

**Definition 13.27.** The set $\mathbb{R}$ is the quotient set $\mathscr{C} / \sim$, where $\sim$ is defined as in Definition 13.25. We equip $\mathbb{R}$ with operations of addition and multiplication as follows:

$$[(x_n)] + [(y_n)] = [(x_n + y_n)],$$
$$[(x_n)] \times [(y_n)] = [(x_n y_n)].$$

**Proposition 13.28.** *The set $\mathbb{R}$, equipped with the operations $+, \times$ just defined, is a field.*

**Proof.** We should first check that the operations $+, \times$ are well-defined, that is, independent of the (arbitrary) choices of representatives of $[(x_n)]$, $[(y_n)]$ used to define them. So, let $(x'_n) \sim (x_n)$ and $(y'_n) \sim (y_n)$. Then $x'_n = x_n + h_n$ and $y'_n = y_n + k_n$ where $(h_n)$ and $(k_n)$ are null. Hence $x'_n + y'_n = x_n + y_n + h_n + k_n$, and $(h_n + k_n)$ is null, so $(x'_n + y'_n) \sim (x_n + y_n)$. It follows that $[(x_n)] + [(y_n)]$ is well-defined. Similarly, $x'_n y'_n = x_n y_n + x_n k_n + y_n h_n + h_n k_n$. Now, since $(x_n), (y_n)$ are Cauchy, they are bounded (Lemma 13.18), and hence $(x_n k_n)$ and $(y_n h_n)$ are null (see Tutorial Problem 2 in Chapter 3, page 43). It follows that $(x_n k_n + y_n h_n + h_n k_n)$ is null (Algebra of Limits), and hence $(x'_n y'_n) \sim (x_n y_n)$. Hence $[(x_n)] \times [(y_n)]$ is well-defined. It is clear that both $+$ and $\times$ are commutative and associative and that $\times$ distributes over $+$ (since $+, \times$ on $\mathbb{Q}$ have these properties).

We must now check that $(\mathbb{R}, +, \times)$ satisfies the other axioms of a field (Definition 13.10). There is an additive inverse, namely the equivalence class of the constant sequence $(0)$. More compactly, $0_{\mathbb{R}} = [(0)] = \mathscr{N}$, the set of null sequences. There is also a multiplicative identity, namely the class of the constant sequence $(1)$, that is, $1_{\mathbb{R}} = [(1)]$. The additive inverse of $[(x_n)]$ is $[(-x_n)]$. Defining multiplicative inverses is a bit trickier. Let $[(x_n)] \in \mathbb{R}$ and $[(x_n)] \neq 0_{\mathbb{R}}$. This means precisely that $(x_n) \notin \mathscr{N}$. But then, by Proposition 13.23, $(x_n) \in \mathscr{C}_+$ or $(x_n) \in \mathscr{C}_-$. In either case, there exists $N \in \mathbb{Z}^+$ and $\varepsilon \in \mathbb{Q}_+$ such that for all $n \geq N$, $|x_n| > \varepsilon$. In particular, for all $n \geq N$, $x_n \neq 0$. Consider the sequence

$$h_n = \begin{cases} 1 \text{ if } x_n = 0 \\ 0 \text{ if } x_n \neq 0 \end{cases}.$$

As just argued, $h_n = 0$ for all $n \geq N$, so $(h_n)$ is certainly null. Now, by construction, $(x'_n) = (x_n + h_n)$ is always non-zero and differs from $(x_n)$ by a null sequence, so $(x'_n) \in [(x_n)]$. We define

$$[(x_n)]^{-1} = [(1/x'_n)].$$

We leave it as an exercise for the reader to check that $(1/x'_n)$ is a Cauchy sequence, and that its equivalence class is independent of the choices of $(x_n)$ in $[(x_n)]$, $N \in \mathbb{Z}^+$, and $\varepsilon \in \mathbb{Q}_+$ used to define it. $\qquad \square$

**Remark.** If you know a bit more advanced algebra, the proof above can be shortened (and smartened up) considerably. One notes that $\mathscr{C}$ equipped with termwise addition and multiplication is a commutative *ring* with multiplicative identity, and that $\mathscr{N} \subset \mathscr{C}$ is an *ideal*. Then $(\mathbb{R}, +, \times)$ is precisely the *quotient ring* $\mathscr{C}/\mathscr{N}$, which is automatically a field since the ideal $\mathscr{N}$ is *maximal* (showing this is really the only nontrivial part). This line of argument is certainly elegant, but it requires one to develop a large body of algebraic theory (for example, defining all the italicized terms just mentioned) which would be rather out of place here.

So we have a constructive definition of $\mathbb{R}$ as a field. We must next equip it with an ordering relation. Recall we can do this by specifying the subset of $\mathbb{R}$ consisting of "positive" elements (see Definition 13.12).

**Definition 13.29.** An element $[(x_n)] \in \mathbb{R}$ is said to be **positive** if $(x_n) \in \mathscr{C}_+$, that is, if $(x_n)$ is strongly positive. Note that this condition is independent of the choice of representative sequence $(x_n)$ by Proposition 13.24. We denote the set of positive reals $\mathbb{R}_+$.

See Definition 13.22 to recall what a "strongly positive" rational Cauchy sequence is.

**Proposition 13.30.** $(\mathbb{R}, \mathbb{R}_+)$ *is an ordered field. That is:*

*(i) for all* $[(x_n)], [(y_n)] \in \mathbb{R}_+$, $[(x_n)] + [(y_n)] \in \mathbb{R}_+$ *and* $[(x_n)] \times [(y_n)] \in \mathbb{R}_+$;

*(ii) for all $[(x_n)] \in \mathbb{R}$, exactly one of $[(x_n)] \in \mathbb{R}_+$, $[(x_n)] = 0_\mathbb{R}$, $-[(x_n)] \in \mathbb{R}_+$ is true.*

**Proof.** (i) This amounts to the statement that $\mathscr{C}_+$ is closed under addition and multiplication, which follows easily from Definition 13.22 and Proposition 13.19.

(ii) This follows directly from Proposition 13.23.

$\square$

Since $(\mathbb{R}, \mathbb{R}_+)$ is an ordered field, it has a natural ordering relation, and all the definitions in section 1.3 make sense. So we know what it means for a subset $A$ of $\mathbb{R}$ to be bounded above (and/or below), and we know what the *supremum* of such a set is: it is the smallest of all the upper bounds on $A$ in $\mathbb{R}$ – *if* such an element exists! Our final task is to show that, for all nonempty subsets $A$ of $\mathbb{R}$ that are bounded above, the supremum *does* exist. Having *constructed* $\mathbb{R}$ (as a set of equivalence classes of rational Cauchy sequences), we will then have proved that it satisfies the Axiom of Completeness and thus supports the whole body of theory (convergence, continuity, and differential and integral calculus) developed in the preceding chapters. This, then, is a big one:

**Theorem 13.31 (Completeness of $\mathbb{R}$).** *Let $A \subset \mathbb{R}$ be nonempty and bounded above. Then $A$ has a supremum in $\mathbb{R}$.*

The proof will appeal several times to another fundamental property of $\mathbb{R}$, the *Archimedean Property*. Recall that in Chapter 1 we proved that the Archimedean Property follows from the Axiom of Completeness. Of course, we can't use that result here since our goal is to prove that $\mathbb{R}$ satisfies the Axiom of Completeness, so our logic would be neatly circular. So, we must give an independent proof that $\mathbb{R}$, as we have constructed it, has the Archimedean Property. To do this, we should first note that $\mathbb{R}$ contains embedded within it a copy of $\mathbb{Q}$, and within that a copy of $\mathbb{Z}^+$. We just identify $q \in \mathbb{Q}$ with the equivalence class of the constant sequence $(q)$.

**Theorem 13.32 (Archimedean Property of $\mathbb{R}$).** *The subset $\mathbb{Z}^+ \subset \mathbb{R}$ is unbounded above. That is, for all $x \in \mathbb{R}$, there exists $k \in \mathbb{Z}^+$ such that $k > x$.*

**Proof.** Let $x \in \mathbb{R}$. Then $x = [(x_n)]$ for some rational Cauchy sequence $(x_n)$. Since $(x_n)$ is Cauchy, it is bounded (Lemma 13.18), that is, there exists $q \in \mathbb{Q}$ such that, for all $n \in \mathbb{Z}^+$, $x_n < q$. Now $\mathbb{Q}$ has the Archimedean

Property, so there exists $k' \in \mathbb{Z}^+$ such that $q < k'$. Let $k = k' + 1$. Then, for all $n \in \mathbb{Z}^+$, $x_n < k - 1$. We identify $k$ with the real number $k_{\mathbb{R}} = [(k, k, k, \ldots)]$. Then, for all $n \in \mathbb{Z}^+$, $k - x_n > 1$, so certainly $(k - x_n) \in \mathscr{C}_+$, whence $k_{\mathbb{R}} - x \in \mathbb{R}_+$. Hence $k_{\mathbb{R}} > x$, as required.                     $\square$

*Proof of Theorem 13.31:*   Let $A \subset \mathbb{R}$ be nonempty and bounded above, by $K \in \mathbb{R}$ say. By the Archimedean Property, there exists $M \in \mathbb{Z}^+$ such that $M > K$. Then certainly $M$ is an upper bound on $A$ (since $<$ is transitive). Since $A$ is nonempty, there exists $y \in A$. By the Archimedean Property, there exists $m' \in \mathbb{Z}^+$ such that $m' > -y$. Hence, there exists $m \in \mathbb{Z}$ (namely $m = -m'$) such that $m < y$. Since $y \in A$, it follows that $m$ is *not* an upper bound on $A$. Now, for each $n \in \mathbb{Z}^+$, let

$$ Q_n = \{ \frac{k}{2^n} \ : \ k \in \mathbb{Z}, \ m \leq \frac{k}{2^n} \leq M \}. $$

Note that this is a finite subset of $\mathbb{Q}$, that for all $n$, $Q_n$ contains $M$ (just take $k = 2^n M \in \mathbb{Z}$) and that, for all $n$, $Q_n \subseteq Q_{n+1}$ (if $m \leq k/2^n \leq M$, then $m \leq (2k)/2^{n+1} \leq M$). Further, for each $n \in \mathbb{Z}^+$, let

$$ B_n = \{ q \in Q_n \ : \ q \text{ is an upper bound on } A \}. $$

Each $B_n$ is nonempty, since it contains the upper bound $M$, and is finite, since it is a subset of the finite set $Q_n$. Also, for all $n \in \mathbb{Z}^+$, $B_n \subseteq B_{n+1}$ (since $Q_n \subseteq Q_{n+1}$). Since $B_n$ is finite, it must have a smallest element, call it $a_n \in \mathbb{Q}$. Consider the rational sequence $(a_n)$. Since $B_{n+1} \supseteq B_n$, $a_{n+1} \leq a_n$ (the set $B_{n+1}$ contains $B_n$ so its minimum element can't be larger than the minimum element of $B_n$). That is, the sequence $(a_n)$ is decreasing. Clearly $a_n - \frac{1}{2^n}$ is *not* an upper bound on $A$ (either it is in $Q_n$, in which case it can't be an upper bound on $A$ since $a_n$ is by definition the *least* upper bound on $A$ in $Q_n$; or it is not in $Q_n$, in which case, since $a_n$ *is* in $Q_n$, $a_n - \frac{1}{2^n} < m$ and hence, being smaller than $m$, which is *not* an upper bound on $A$, likewise cannot be an upper bound on $A$). On the other hand, $a_{n+1}$ *is* an upper bound on $A$, so we deduce that

$$ a_{n+1} > a_n - \frac{1}{2^n}. $$

Together with the fact that $(a_n)$ is decreasing, it follows that, for all $n \in \mathbb{Z}^+$,

$$ |a_{n+1} - a_n| = a_n - a_{n+1} < \frac{1}{2^n}. $$

Hence, by Lemma 13.21, $(a_n)$ is Cauchy. Denote by $a$ the corresponding real number, that is, equivalence class $[(a_n)]$. We claim that $a$ is the supremum of $A$, that is, $a$ is an upper bound on $A$ and every real number less than $a$ is *not* an upper bound on $A$. It remains to establish these two facts.

*Claim 1:* $a$ is an upper bound on $A$.

Assume this is false. Then there exists $x \in A$ such that $x > a$. By the Archimedean Property of $\mathbb{R}$, there exists $N \in \mathbb{Z}^+$ such that $N > 1/(x-a)$. Hence

$$x - a > \frac{1}{N} > \frac{1}{2^N}. \tag{13.1}$$

Now, for all $n, m \in \mathbb{Z}^+$, $a_n$ is an upper bound on $A$ while $a_m - \frac{1}{2^m}$ is not, so

$$a_n > a_m - \frac{1}{2^m}.$$

Hence, for each *fixed* $m \in \mathbb{Z}^+$, the sequence

$$\left(a_n - \left(a_m - \frac{1}{2^m}\right)\right)$$

has only positive terms, and hence is certainly *not* in $\mathscr{C}_-$. Hence, this sequence is in $\mathscr{N}$ or $\mathscr{C}_+$. Consider now the real numbers defined by $(a_n)$ and the constant sequence taking value $a_m - \frac{1}{2^m}$, namely $a$ and $a_m - \frac{1}{2^m}$. Since the difference of their representative sequences is not in $\mathscr{C}_-$, it follows that $a - (a_m - \frac{1}{2^m}) \geq 0$, that is

$$a \geq a_m - \frac{1}{2^m}.$$

This is true for every value of $m \in \mathbb{Z}^+$, including the positive integer $N$ introduced above. Combining this with (13.1) above, we see that

$$x > a + \frac{1}{2^N} \geq a_N - \frac{1}{2^N} + \frac{1}{2^N} = a_N.$$

But this contradicts the definition of $a_N$ which was, recall, an upper bound on $A$. Hence Claim 1 is established.

*Claim 2:* If $b < a$ then $b$ is *not* an upper bound on $A$.

Let $b \in \mathbb{R}$ such that $b < a$. By the Archimedean Property of $\mathbb{R}$, there exists $N \in \mathbb{Z}^+$ such that $N > 1/(a-b)$, and hence

$$a - b > \frac{1}{N} > \frac{1}{2^N} \quad \Rightarrow \quad a - \frac{1}{2^N} > b. \tag{13.2}$$

Since $(a_n)$ is decreasing, $a_n - a_N \leq 0$ for all $n \geq N$, and hence the sequence $(a_n - a_N)$ is eventually non-positive, so clearly cannot be in $\mathscr{C}_+$. It must therefore be in $\mathscr{C}_-$ or $\mathscr{N}$. Hence, comparing the real numbers $a$ (the class of $(a_n)$) and $a_N$ (the class of the constant sequence $(a_N, a_N, a_N, \ldots)$) we see that $a \leq a_N$. But then, from (13.2), one sees that

$$b < a_N - \frac{1}{2^N}.$$

As we have previously argued, $a_N - \frac{1}{2^N}$ is *not* an upper bound on $A$, so $b$, being smaller still, is likewise not an upper bound on $A$.

$\square$

# Epilogue: let there be $\delta$

It is time for us to confess that the approach to real analysis we have taken in this book is something of a minority sport. Recall (Definition 8.1) that, for us, $\lim_{x \to a} f(x) = L$ means that, for all sequences $(x_n)$ in the domain of $f$ which converge to, but never equal, $a$, $f(x_n) \to L$. The more usual approach is to define limits of functions as follows:

**Definition Ep.1.** Let $D \subseteq \mathbb{R}$, $f : D \to \mathbb{R}$, $a$ be a cluster point of $f$ and $L \in \mathbb{R}$. Then $f$ has **(conventional) limit** $L$ at $a$ if, for all $\varepsilon > 0$, there exists $\delta > 0$ such that, for all $x \in D$, if $0 < |x - a| < \delta$, then $|f(x) - L| < \varepsilon$. Let us temporarily denote this statement by

$$\lim_{x \to a}{}' f(x) = L.$$

(We will see shortly that this definition of limit is actually equivalent to ours, so the prime on $\lim'$ can be discarded.)

This is often called the $\varepsilon$–$\delta$ definition of limit. Actually, it contains the term *cluster point*, which we also gave a sequential definition (Definition 8.2): $a \in \mathbb{R}$ is a **cluster point** of $D \subseteq \mathbb{R}$ if there exists a sequence in $D \backslash \{a\}$ that converges to $a$. Again, it is more usual to define this without reference to sequences:

**Definition Ep.2.** A number $a \in \mathbb{R}$ is a **(conventional) cluster point** of $D \subseteq \mathbb{R}$ if, for all $\varepsilon > 0$, there exists $x \in D$ such that $0 < |x - a| < \varepsilon$.

Let's show that these two definitions of cluster point are equivalent.

**Proposition Ep.3.** *Let $D \subseteq \mathbb{R}$ and $a \in \mathbb{R}$. Then $a$ is a conventional cluster point of $D$ if and only if $a$ is a cluster point of $D$.*

**Proof.** Assume $a$ is a conventional cluster point of $D$. Then, for each $n \in \mathbb{Z}^+$, $\varepsilon = 1/n$ is a positive real number, so there exists $x_n \in D$ such that $0 < |x_n - a| < \varepsilon$, whence $x_n \neq a$ (since $|x_n - a| > 0$) and $a - 1/n < x_n < a + 1/n$ (since $|x_n - a| < \varepsilon$). Consider the sequence $(x_n)$. It lies in $D \backslash \{a\}$ and converges to $a$ (by the Squeeze Rule). Hence, $a$ is a cluster point of $D$.

Conversely, assume $a$ is a cluster point of $D$. Let $\varepsilon > 0$ be given. By assumption, there exists a sequence $(x_n)$ in $D \backslash \{a\}$ with $x_n \to a$. Hence, by the definition of convergence, there exists $N \in \mathbb{Z}^+$ such that for all $n \geq N$, $|x_n - a| < \varepsilon$. In particular, $x_N \neq a$ and $|x_N - a| < \varepsilon$, so $0 < |x_N - a| < \varepsilon$. Such a number $x_N \in D$ exists for any given $\varepsilon > 0$, so $a$ is a conventional cluster point of $D$. $\qquad\square$

Rather similar arguments show that the two definitions of limit (conventional and sequential) are exactly equivalent:

**Proposition Ep.4.** $\lim_{x \to a} f(x) = L$ *if and only if* $\lim'_{x \to a} f(x) = L$.

**Proof.** <u>If direction</u>: assume that $\lim'_{x \to a} f(x) = L$. Let $(x_n)$ be any sequence in $D \backslash \{a\}$ such that $x_n \to a$. We must show that $f(x_n) \to L$. So, let $\varepsilon > 0$ be given. By the assumed property of $f$, there exists $\delta > 0$ such that, whenever $0 < |x - a| < \delta$, $|f(x) - L| < \varepsilon$. Choose such a $\delta$. Since $\delta > 0$, and $x_n \to a$, there exists $N \in \mathbb{Z}^+$ such that, for all $n \geq N$, $|x_n - a| < \delta$. But $x_n \neq a$, so for all $n \geq N$, $0 < |x_n - a| < \delta$. But then $|f(x_n) - L| < \varepsilon$ (by the definition of $\delta$). So, we have shown that, given any $\varepsilon > 0$, there exists $N \in \mathbb{Z}^+$ such that, for all $n \geq N$, $|f(x_n) - L| < \varepsilon$. Hence $f(x_n) \to L$, so $\lim_{x \to a} f(x) = L$.

<u>Only if direction</u>: we will prove the *contrapositive*, that is, we will prove that if $\lim'_{x \to a} f(x) \neq L$, then $\lim_{x \to a} f(x) \neq L$. So, assume it is *false* that, for all $\varepsilon > 0$, there exists $\delta > 0$ such that, for all $x \in [D \cap (a - \delta, a + \delta)] \backslash \{a\}$, $|f(x) - L| < \varepsilon$. Then there exists $\varepsilon > 0$ such that, for all $\delta > 0$, there is some $x \in [D \cap (a - \delta, a + \delta)] \backslash \{a\}$ such that $|f(x) - L| \geq \varepsilon$. Choose and fix such a number $\varepsilon$. Then in the particular case $\delta = 1/n$, where $n \in \mathbb{Z}^+$, there exists $x_n \in [D \cap (a - 1/n, a + 1/n)] \backslash \{a\}$ such that $|f(x_n) - L| \geq \varepsilon$. Consider the sequence $(x_n)$. It is a sequence in $D \backslash \{a\}$ and $a - 1/n < x_n < a + 1/n$, so $x_n \to a$. But, for all $n$, $|f(x_n) - L| \geq \varepsilon$, so $f(x_n)$ certainly does not converge to $L$. Since we have exhibited a sequence in $D \backslash \{a\}$ converging to $a$ whose image under $f$ does not converge to $L$, it follows that $\lim_{x \to a} f(x) \neq L$. $\qquad\square$

So we can drop the term "conventional" from our use of cluster point and limit (and remove the prime from lim$'$) without ambiguity.

Having defined limits of functions without reference to sequences, it's usual to use them to define *continuity*:

**Definition Ep.5.** Let $D \subseteq \mathbb{R}$, $f : D \to \mathbb{R}$ and $a \in D$. Then $f$ is **(conventionally) continuous** at $a$ if $a$ is not a cluster point of $D$ or $\lim_{x \to a} f(x) = f(a)$.

By contrast, we defined continuity as follows (Definition 6.1): $f : D \to \mathbb{R}$ is **continuous** at $a \in D$ if, for every sequence $(x_n)$ in $D$ converging to $a$, $f(x_n) \to f(a)$. We have actually already proved that this is equivalent to Definition Ep.5: see Proposition 8.10.

Broadly speaking, the difference between our approach and the more conventional one is this. Conventionally, one makes "$\varepsilon$–$\delta$" definitions of the key objects in analysis (limits, continuous functions etc.), similar to Definition Ep.1, then (if desired) proves that these have equivalent "sequential" characterizations, which can be used when convenient. The underlying concepts themselves are not thought of in a sequential way. Our approach has been to stand this on its head. We *define* all the fundamental objects of analysis using sequences and consistently think of them in sequential fashion. We *could* have proved that they have equivalent "$\varepsilon$–$\delta$" style characterizations had we wanted to (see Propositions Ep.3 and Ep.4), but actually we have no need of these, so we didn't bother. Indeed, the only reason to include Propositions Ep.3 and Ep.4 at all is to point out that our approach *does* differ from the usual one, allow comparison of the two, and assure the reader that, although the definitions, theorems, and proofs contained herein may *look* different from those in other undergraduate texts on real analysis, they are equivalent and develop the same rigorous underpinning for calculus.

# Further reading

- Applebaum, D. (2012). *Limits, Limits Everywhere. The Tools of Mathematical Analysis* (Oxford University Press, Oxford, UK).

  A very detailed and discursive account of (sequential) limits covering a variety of interesting topics in pure mathematics (Fibonacci numbers, irrationality of $\pi$, and continued fractions, for example), this book stops short of developing calculus. It sketches some more advanced ideas, such as ordinal numbers and the construction of $\mathbb{R}$ by Dedekind cuts.

- Bartle, R. G. and Sherbert, D. R. (2011). *Introduction to Real Analysis* (Wiley, Hoboken, NJ, USA).

  A comprehensive account of real analysis, pitched at a somewhat higher level than I've aimed at here, this was the text from which I first learned analysis myself. It develops extensively both sequential and $\varepsilon$–$\delta$ notions, and their interaction, and ends with some basic ideas from topology. Its approach to integration theory is rather different from ours. It does not attempt a constructive definition of the reals.

- Clark, C. W. (1982). *Elementary Mathematical Analysis* (Wadsworth, Canada).

  A more conventional analysis text aimed at undergraduates, this concentrates on $\varepsilon$–$\delta$ ideas and develops some notions of limit that we have neglected (one-sided limits, for example).

- Haggarty, R. (1989). *Fundamentals of Mathematical Analysis* (Addison-Wesley, Wokingham, UK).

  Another conventional introductory text, this has a particularly good section on Taylor's Theorem, developing several interesting alternative forms.

- Priestley, H. A. (1985) *Introduction to Complex Analysis* (Clarendon Press, Oxford, UK).

  Power series and the idea of differentiability become immensely more powerful once one moves from the real line to the complex plane. So, of all developments from real analysis, complex analysis is the most elegant and powerful. This book is a clear but concise introduction to the subject that includes a variety of interesting applications (including a beautiful proof that

  $$\sum_{n=1}^{\infty} \frac{1}{n^2} = \frac{\pi^2}{6},$$

  for example).

- Thurston, H. (1988). *Intermediate Mathematical Analysis* (Oxford University Press, Oxford, UK).

  A rigorous introduction to differential and integral calculus in higher dimensions (i.e. for functions $f : \mathbb{R}^n \to \mathbb{R}^m$ with $n \geq 2$).

# Solutions to tutorial problems

1. $A = (-\infty, -3) \cup [0, 5) \cup [7, \infty)$.
2. (a)   • Let $x_1, x_2 \in \mathbb{R}\backslash\{-1\}$. Then

$$f(x_1) = f(x_2) \Rightarrow \frac{1}{x_1 + 1} = \frac{1}{x_2 + 1}$$
$$\Rightarrow x_2 + 1 = x_1 + 1$$
$$\Rightarrow x_2 = x_1.$$

Hence $f$ is injective.

   • Let $y \in \mathbb{R}\backslash\{0\}$. Then $x = -1 + \dfrac{1}{y} \in \mathbb{R}\backslash\{-1\}$ and

$$f(x) = \frac{1}{(-1 + \frac{1}{y}) + 1} = y.$$

Hence $y$ is in the range of $f$. Hence $f$ is surjective. Hence $f$ is bijective.

*Remark: of course, I found the $x$ whose image under $f$ is $y$ by solving the equation $f(x) = y$ for $x$ as a function of $y$.*

   (b) $0 \in \mathbb{R}$ and $1 \in \mathbb{R}$ and $0 \neq 1$, but $g(0) = g(1) = 0$. Hence $g$ is not injective.

3. (a) Let $y \in B$. Since $f$ is surjective, there exists $x \in A$ such that $f(x) = y$. Then $g(f(x)) = g(y)$, so $x = g(y)$ (since $g \circ f = \text{Id}_A$). Hence $f(g(y)) = f(x) = y$, by the definition of $x$. So $f \circ g = \text{Id}_B$, as claimed.

   (b) Let $x \in A$. Then $f(g(f(x))) = \text{Id}_B(f(x)) = f(x)$. But $f$ is injective, and $f((g \circ f)(x)) = f(x)$, so $(g \circ f)(x) = x$. Hence $g \circ f = \text{Id}_A$.

4.   • $A$ is bounded below by $-2$.

227

**Proof.** Let $z \in A$. Then there exist $x \in \mathbb{R}$ and $y \in [-1, 1]$ such that $z = x^2 + 2x + y$. But then

$$z = (x+1)^2 - 1 + y \geq -1 + y \qquad (\text{since } (x+1)^2 \geq 0)$$
$$\geq -1 - 1 = -2 \qquad (\text{since } y \geq -1).$$

□

- $A$ is unbounded above.

**Proof.** Clearly $1 \in A$ (take $x = 0$, $y = 1$), so if $K$ is an upper bound on $A$ then $K \geq 1$. But then $K^2 + 2K \in A$ (take $x = K$ and $y = 0$) and $K^2 + 2K > 2K > K$, so $K$ is not an upper bound on $A$. □

5. (a)
    - [1 is an upper bound on $A$] Let $x \in A$. Then there exist $n, m \in \mathbb{Z}^+$ such that $x = n^{-1} - 2m^{-1}$. But then $m > 0$ so $-2m^{-1} < 0$ and hence

    $$x < \frac{1}{n} \leq 1$$

    since $n \geq 1$.
    - [No real number less than 1 is an upper bound on $A$] Let $K < 1$. Then $1 - K > 0$, so $2/(1 - K) > 0$, and hence, by the Archimedean Property, there exists $m \in \mathbb{Z}^+$ such that $m > 2/(1 - K)$. But then $0 < 2/m < 1 - K$, so

    $$1 - \frac{2}{m} > 1 - (1 - K) = K$$

    and $1 - (2/m) \in A$ (take $n = 1$), so $K$ is not an upper bound on $A$.

    (b) No, $1 \notin A$. In fact, we already showed this above, since we showed that every $x \in A$ has $x < 1$ (strict inequality).

    (c) Yes, $A$ is bounded below, so it has an infimum. In fact $\inf A = -2$. It's not hard to prove this using a similar argument to the one above.

# Chapter 2

1. Claim: $a_n \to 2$.

**Proof.** Let $\varepsilon > 0$ be given. Then $5/\varepsilon > 0$ is a real number so, by the Archimedean Property of $\mathbb{R}$, there exists $N \in \mathbb{Z}^+$ such that $N > 5/\varepsilon$. Then for all $n \geq N$,

$$|a_n - 2| = \left| \frac{n+4}{n^2+2} \right| = \frac{n+4}{n^2+2} \leq \frac{5n}{n^2+2} \qquad \text{(since } n \geq 1\text{)}$$

$$< \frac{5n}{n^2} \overset{(*)}{\leq} \frac{5}{N} < \varepsilon.$$

Hence, $a_n \to 2$. $\qquad\square$

*Remark: how did I know that any $N$ bigger than $5/\varepsilon$ would be big enough to force $|a_n - 2| < \varepsilon$ for all $n \geq N$? The answer is that I calculated the chain of inequalities up to $(*)$ before I started to write out the proof!*
Claim: $b_n \to -3$.

**Proof.** Let $\varepsilon > 0$ be given. Then $K = \max\{3, 23/\varepsilon\} > 0$ is a real number so, by the Archimedean Property of $\mathbb{R}$, there exists $N \in \mathbb{Z}^+$ such that $N > K$. Then for all $n \geq N$,

$$|b_n - (-3)| = \left| \frac{1 - (-1)^n + 6n^2 + 21 - 6n^2}{7 - 2n^2} \right|$$

$$\leq \frac{22 + |(-1)^n|}{|7 - 2n^2|} \qquad \text{(Triangle Inequality)}$$

$$= \frac{23}{|n^2 + (n^2 - 7)|} \leq \frac{23}{n^2 + 2} \qquad \text{(since } n \geq 3\text{)}$$

$$< \frac{23}{n^2} \leq \frac{23}{n} \qquad \text{(since } n \geq 1\text{)}$$

$$< \varepsilon \qquad \text{(since } n > 23/\varepsilon\text{)}.$$

Hence $b_n \to -3$. $\qquad\square$

Claim: $c_n \to 0$.

**Proof.** Let $\varepsilon > 0$ be given. By the Archimedean Property of $\mathbb{R}$, there exists $N \in \mathbb{Z}^+$ such that $N > 1/\varepsilon^2$. Then for all $n \geq N$,

$$|c_n - 0| = \frac{(\sqrt{n+1} - \sqrt{n})(\sqrt{n+1} + \sqrt{n})}{\sqrt{n+1} + \sqrt{n}} = \frac{1}{\sqrt{n+1} + \sqrt{n}} < \frac{1}{\sqrt{n}} \leq \frac{1}{\sqrt{N}} < \varepsilon.$$

Hence $c_n \to 0$. $\qquad\square$

Claim: $d_n \to 0$.

**Proof.** First, we claim that, for all $n \geq 2$, $d_n \leq 4/n$. We establish this by induction. If $n = 2$ then $d_n = 2^2/2! = 4/2 = 4/n$, so the claim holds for $n = 2$. Assume the claim holds for some $n = k \geq 2$. Then

$$d_{k+1} = \frac{2^{k+1}}{(k+1)!} = \frac{2}{k+1}d_k \leq \frac{2}{k+1}\frac{4}{k} \leq \frac{4}{k+1} \quad \text{(since } k \geq 2\text{)},$$

so it also holds for $n = k + 1$. Hence, by induction, $d_n \leq 4/n$ for all $n \geq 2$.

Now, let $\varepsilon > 0$ be given. By the Archimedean Property of $\mathbb{R}$, there exists $N \in \mathbb{Z}^+$ such that $N > \max\{1, 4/\varepsilon\}$. Then for all $n \geq N$,

$$|d_n - 0| = d_n \leq \frac{4}{n} \quad \text{(since } n \geq 2\text{)}$$
$$< \varepsilon \quad \text{(since } n \geq 4/\varepsilon\text{)}.$$

Hence $d_n \to 0$. $\qquad\qquad\qquad\qquad\qquad\qquad\qquad\qquad\qquad\qquad\quad\square$

2.  • Clearly $(1 + \beta)^1 = 1 + \beta \geq 1 + 1 \times \beta$, so the claim holds for $n = 1$. Assume the claim holds for $n = k$. Then

$$(1+\beta)^{k+1} = (1+\beta)(1+\beta)^k \geq (1+\beta)(1+k\beta) = 1+(k+1)\beta+\beta^2 > 1+(k+1)\beta$$

so the claim also holds for $n = k+1$. Hence, by induction, the claim holds for all $n \in \mathbb{Z}^+$.

  • Let $\varepsilon > 0$ be given. By the Archimedean Property of $\mathbb{R}$, there exists $N \in \mathbb{Z}^+$ such that $N > 1/(\varepsilon\beta)$. Then for all $n \geq N$,

$$|a_n - 0| = a_n \leq \frac{1}{1+n\beta} < \frac{1}{n\beta} \leq \frac{1}{N\beta} < \varepsilon.$$

Hence $a_n \to 0$.

## Chapter 3

1. (a) $a_n = \dfrac{7 + (2/n^2)}{2 + (11/n)} \to \dfrac{7+0}{2+0}$ by Theorem 3.5.

   (b) $\dfrac{-1}{n^2} \leq b_n \leq \dfrac{1}{n^2}$ for all $n$, and $-1/n^2 \to 0$ and $1/n^2 \to 0$. Hence $b_n \to 0$ by the Squeeze Rule.

   (c) $\dfrac{c_{n+1}}{c_n} = \dfrac{n+1}{2n} = \dfrac{n+1}{n+n} \leq 1$ since $n \geq 1$. Hence $c_{n+1} \leq c_n$ for all $n$, that is, $(c_n)$ is decreasing. Clearly $c_n > 0$ for all $n$, so $(c_n)$ is bounded below. Hence, by the Monotone Convergence Theorem, $(c_n)$ converges.

*Remark: the question doesn't ask you to determine the sequence's limit, so the above proof suffices. But while we're here, why don't we try to figure it out?*

Note that

$$c_{n+1} = \frac{n+1}{2^{n+1}} = \frac{1}{2}\left(\frac{n}{2^n} + \frac{1}{2^n}\right) = \frac{1}{2}\left(c_n + \frac{1}{2^n}\right).$$

Now, assume that $c_n \to L$. Then "clearly" $c_{n+1} \to L$ also (actually this follows from Theorem 4.3). Furthermore, $1/2^n = (0.5)^n \to 0$ by Example 3.15. Hence, by the Algebra of Limits, $c_{n+1} \to \frac{1}{2}(L+0)$. But $\lim c_{n+1}$ is unique, so $L = L/2$, and hence $L = 0$. That is, $c_n \to 0$.

2. (a) Let $\varepsilon > 0$ be given. Since $(a_n)$ is bounded, there exists $K > 0$ such that $|a_n| \le K$ for all $n$. Now $\varepsilon' = \varepsilon/K > 0$, and $b_n \to 0$, so there exists $N \in \mathbb{Z}^+$ such that, for all $n \ge N$, $|b_n - 0| < \varepsilon'$. But then, for all $n \ge N$,

$$|a_n b_n - 0| = |a_n||b_n| \le K|b_n| < K\varepsilon' = \varepsilon.$$

Hence, $a_n b_n \to 0$.

(b) No. The sequences $a_n = n$, $b_n = 0$ give a counterexample ($a_n$ is unbounded above, but $a_n b_n = 0 \to 0$).

(c) No. The sequences

$$a_n = \begin{cases} 0 & n \text{ odd} \\ 1 & n \text{ even} \end{cases}, \qquad b_n = \begin{cases} n & n \text{ odd} \\ 0 & n \text{ even} \end{cases}$$

give a counterexample ($a_n$ is bounded but not convergent, $a_n b_n = 0 \to 0$, but $b_n$ is unbounded, hence certainly doesn't converge to 0).

# Chapter 4

1. (a) i. The first few terms of this sequence are

$$(a_n) = (-2/7, 0, -2/3, 0, 2, 0, 2/5, 0, 2/9, \ldots).$$

From this we see that the $a_1, a_2, a_3, a_4$ are certainly *not* dominant, since they are each smaller than $a_5$. In fact, $a_n = 0$ for all even $n$ and for all *odd* $n \ge 5$, $a_n = 2/(2n-9) > 0$, so no even numbered term is dominant either. Consider the terms $a_{2k+3}$ where $k \in \mathbb{Z}^+$,

$$a_{2k+3} = \frac{2}{4k-3},$$

which is a positive decreasing sequence. Hence each of these terms is dominant, that is, the set of dominant terms is $\{a_5, a_7, a_9, \ldots\}$.

ii. Since the set of dominant terms is infinite, they form a decreasing subsequence. Explicitly, $a_{2k+3} = 2/(4k-3)$ is a monotone subsequence. Alternatively, $a_{2k} = 0$ is also, strictly speaking, a monotone subsequence (it's both increasing and decreasing!).

iii. Yes, $(a_n)$ does have a convergent subsequence, e.g. the monotone sequence constructed above. In fact, $(a_n)$ itself is convergent, so *every* subsequence of $(a_n)$ is convergent (by Theorem 4.3).

(b)  i. Note that

$$b_n = \begin{cases} -3n & n \text{ odd} \\ -n & n \text{ even} \end{cases}.$$

Hence, no odd term is dominant, since the following even term is bigger ($a_{2k+1} = -3(2k+1) = -6k - 3 < -2k - 2 = a_{2k+2}$). On the other hand, every even term is dominant since if $m$ is even and $n \geq m$ then $a_n \leq -n \leq -m = a_m$. So the set of dominant terms is $\{a_2, a_4, a_6, \ldots\}$.

ii. Again, the set of dominant terms, being infinite, forms a decreasing subsequence. Alternatively, $a_{2k-1} = -3(2k-1)$ is also a decreasing subsequence (consisting of all odd terms of $(a_n)$).

iii. In this case, for any subsequence $a_{n_k} \leq -n_k \leq -k$, so every subsequence is unbounded below. Hence, no, this sequence has no convergent subsequences.

2. Since $(a_n)$ is unbounded, for each $k \in \mathbb{Z}^+$ there is some $n_k \in \mathbb{Z}^+$ such that $|a_{n_k}| > k$. Choose such an $n_k$ and consider the subsequence $(a_{n_k})$. I claim that $1/a_{n_k} \to 0$. Proof: let $\varepsilon > 0$ be given. Then, by the Archimedean Property of $\mathbb{R}$, there exists $K \in \mathbb{Z}^+$ such that $K > 1/\varepsilon$. Now for all $k \geq K$,

$$\left| \frac{1}{a_{n_k}} - 0 \right| = \frac{1}{|a_{n_k}|} < \frac{1}{k} \leq \frac{1}{K} < \varepsilon.$$

$\square$

3. (i) We note that

$$x_{n+2} = 1 + \frac{1}{x_{n+1}} = 1 + \frac{1}{1 + 1/x_n} = \frac{2x_n + 1}{x_n + 1}, \qquad \text{(T.1)}$$

whence

$$1 + x_{n+2} - x_{n+2}^2 = 1 + \frac{2x_n + 1}{x_n + 1} - \left( \frac{2x_n + 1}{x_n + 1} \right)^2 = \frac{1 + x_n - x_n^2}{(x_n + 1)^2}.$$

Now $1 + x_1 - x_1^2 = 1 > 0$, and, as shown above, if $1 + x_n - x_n^2 > 0$ then $1 + x_{n+2} - x_{n+2}^2 > 0$. Hence, by induction, $1 + x_n - x_n^2 > 0$ for all odd $n$. Similarly, $1 + x_2 - x_2^2 = -1 < 0$, and, as shown above, if $1 + x_n - x_n^2 < 0$ then $1 + x_{n+2} - x_{n+2}^2 < 0$. Hence, by induction, $1 + x_n - x_n^2 < 0$ for all even $n$.

(ii)

$$x_{n+2} - x_n = \frac{2x_n + 1}{x_n + 1} - x_n = \frac{1 + x_n - x_n^2}{x_n + 1}$$

so it follows from part (i) that $(x_{2k-1})$ is increasing and $(x_{2k})$ is decreasing.

(iii) If $x_n \in [1, 2]$ then $\frac{1}{2} \leq 1/x_n \leq 1$ and so $x_{n+1} \in [\frac{3}{2}, 2] \subset [1, 2]$. $x_1 = 1 \in [1, 2]$ so, by induction, $x_n \in [1, 2]$ for all $n$. Hence $(x_n)$ is bounded above and below, so the monotone subsequences $(x_{2k-1})$, $(x_{2k})$ are likewise bounded, and hence, by the Monotone Convergence Theorem, converge.

(iv) Let $x_{2k-1} \to L_1$ and $x_{2k} \to L_2$. Then $x_{2k+1} \to L_1$ and $x_{2k+2} \to L_2$ also, since these are subsequences of $(x_{2k-1})$ and $(x_{2k})$. But, by equation (T.1),

$$x_{2k+1} = \frac{2x_{2k-1} + 1}{x_{2k-1} + 1} \quad \text{and} \quad x_{2k+2} = \frac{2x_{2k} + 1}{x_{2k} + 1}$$

so, by the Algebra of Limits,

$$x_{2k+1} \to \frac{2L_1 + 1}{L_1 + 1} \quad \text{and} \quad x_{2k+2} \to \frac{2L_2 + 1}{L_2 + 1}.$$

But limits are unique, so $L_1 = (2L_1 + 1)/(L_1 + 1)$ and $L_2 = (2L_2 + 1)/(L_2 + 1)$. Hence, both $L_1$ and $L_2$ are roots of the polynomial $L^2 - L - 1$, that is, $(1 \pm \sqrt{5})/2$. It follows that $\sqrt{5}$ exists (i.e. there is a positive real number whose square is 5). Since $a_n \in [1, 2]$ for all $n$, the limits $L_1, L_2 \in [1, 2]$ also, by Proposition 3.4. Clearly $(1 - \sqrt{5})/2 < 1/2 < 1$, so we conclude that

$$L_1 = L_2 = \frac{1}{2}(1 + \sqrt{5}).$$

Now $(x_{2k-1})$ and $(x_{2k})$ form a covering pair for $(x_n)$ so, by Lemma 4.8, $(x_n)$ also converges, to the same limit $(1 + \sqrt{5})/2$.

*Remark: the limit of this sequence is a very famous irrational number called the* **Golden Mean**. *It is (approximately) the ratio of the width of your credit card to its height. This aspect ratio is thought to be particularly aesthetically pleasing and recurs frequently in classical art. Geometrically, it is the aspect ratio of a rectangle with the property that, if one lops off a square from one end, the remaining smaller rectangle,* *rotated through 90°, has the same aspect ratio. The sequence $(x_n)$ we just considered is also interesting, in that it is associated with the **Fibonacci numbers**. These are defined inductively so that $a_1 = a_2 = 1$ and $a_n = a_{n-1} + a_{n-2}$ for all $n \geq 3$. The link is that $x_n = a_{n+1}/a_n$ (check it). So we have just shown that the ratio of consecutive Fibonacci numbers converges to the Golden Mean.*

## Chapter 5

1. (a) We seek constants $A, B, C, D$ such that

$$\frac{n+1}{n^2(n+2)^2} = \frac{A}{n} + \frac{B}{n^2} + \frac{C}{n+2} + \frac{D}{(n+2)^2}$$

$$= \frac{(A+C)n^3 + (4A+B+2C+D)n^2 + (4A+4B)n + 4B}{n^2(n+2)^2}.$$

Comparing coefficients of $n^p$ in the numerator, we see that $A = C = 0$ and $B = -D = 1/4$. Hence

$$a_n = \frac{1}{4}\left(\frac{1}{n^2} - \frac{1}{(n+2)^2}\right),$$

so the $k^{\text{th}}$ partial sum is

$$s_k = \frac{1}{4}\sum_{n=1}^{k}\left(\frac{1}{n^2} - \frac{1}{(n+2)^2}\right) = \frac{1}{4}\left(\sum_{n=1}^{k}\frac{1}{n^2} - \sum_{m=3}^{k+2}\frac{1}{m^2}\right)$$

$$= \frac{1}{4}\left(1 + \frac{1}{4} - \frac{1}{(k+1)^2} - \frac{1}{(k+2)^2}\right).$$

(b) It follows from the Algebra of Limits (Theorem 3.5) that

$$s_k \to \frac{1}{4}\left(1 + \frac{1}{4} - 0 - 0\right) = \frac{5}{16}.$$

2. (a) Claim: the series converges.

   **Proof.** Let $a_n = 2^n/(1+3^n)$. Note that $a_n > 0$ and

$$\frac{a_{n+1}}{a_n} = \frac{2}{(1/3)^n + 3}((1/3)^n + 1) \to \frac{2}{0+3}(0+1) = \frac{2}{3}$$

   by the Algebra of Limits. Since $2/3 < 1$, the series $\sum a_n$ converges by the Ratio Test. $\qquad\square$

   (b) Claim: the series converges.

**Proof.** Let $a_n = n/((n+1)(n+2))$ and note that the series is alternating. Now, for all $n \in \mathbb{Z}^+$,

$$a_n - a_{n+1} = \frac{1}{n+2} \left( \frac{n}{n+1} - \frac{n+1}{n+3} \right) = \frac{n-1}{(n+1)(n+2)(n+3)} \geq 0,$$

so the sequence $(a_n)$ is decreasing. Furthermore,

$$a_n = \frac{(1/n)}{(1+(1/n))(1+(2/n))} \to \frac{0}{(1+0)(1+0)} = 0,$$

by the Algebra of Limits, so the series $\sum (-1)^{n+1} a_n$ converges by the Alternating Series Test. $\qquad \square$

(c) Claim: the series converges.

**Proof.** Let $a_n = (-1)^{n+1}(n+20)^2/n!$. Then $|a_n| > 0$ and

$$\frac{|a_{n+1}|}{|a_n|} = \frac{(n+21)^2}{(n+1)!} \frac{n!}{(n+20)^2} = \frac{1}{n+1} \left( 1 + \frac{1}{n+20} \right)^2 \to 0,$$

by the Algebra of Limits. Hence, the series $\sum |a_n|$ converges by the Ratio Test, that is, the series $\sum a_n$ converges absolutely. Hence, the series $\sum a_n$ converges, by Theorem 5.20. $\qquad \square$

*Remark: this series is alternating too, so we could have tried to use the Alternating Series Test. But showing that $|a_n|$ is a decreasing sequence is rather fiddly, so it's easier just to throw away the extra information given by the alternating sign and prove absolute convergence.*

(d) Claim: the series diverges.

**Proof.** Let $a_n = (8n^2 - 7n + 12)/(4n^3 + n^2 + 5n) > 0$ and $b_n = 1/n > 0$. Then

$$\frac{b_n}{a_n} = \frac{4n^3 + n^2 + 5n}{8n^3 - 7n^2 + 12n} = \frac{4 + (1/n) + (5/n^2)}{8 - (7/n) + (12/n^2)}$$

converges (by the Algebra of Limits), and the series $\sum b_n$ diverges, so $\sum a_n$ diverges by part (ii) of the Comparison Test. $\qquad \square$

*Remark: the idea behind this proof is that, for large $n$, the terms of the series look roughly like $2/n$, so we compare the series with $\sum 1/n$, which we know diverges.*

3. (a) Let $a_n = 3^{-n} \cos(2n\pi/3)$. To show that the series converges absolutely, we must show that the sequence of partial sums

$$t_k = \sum_{n=1}^{k} |a_n|$$

converges. This sequence is increasing, so converges provided it is bounded above (by the Monotone Convergence Theorem). But $|a_n| \leq 1/3^n$, so

$$t_k \leq \sum_{n=1}^{k} \frac{1}{3^n} < \sum_{n=0}^{k} \frac{1}{3^n} = \frac{1 - 1/3^{n+1}}{1 - (1/3)} < \frac{2}{3}.$$

Since $(t_k)$ is bounded above, it converges, so $\sum a_n$ converges absolutely, and hence converges, by Theorem 5.20.

(b) From part (a), we know that the sequence of partial sums

$$s_k = \sum_{n=1}^{k} a_n$$

converges to some limit $L$, say. Hence, every subsequence of $(s_k)$ also converges to $L$ (Theorem 4.3). Consider the subsequence $(s_{3k})$:

$$s_{3k} = a_1 + a_2 + \ldots + a_{3k} = \sum_{n=1}^{k} (a_{3n-2} + a_{3n-1} + a_{3n})$$

$$= \sum_{n=1}^{k} \left( \cos(-4\pi/3) \frac{1}{3^{3n-2}} + \cos(-2\pi/3) \frac{1}{3^{3n-1}} + \cos(0) \frac{1}{3^{3n}} \right)$$

$$= \sum_{n=1}^{k} \frac{1}{27^n} \left( -\frac{1}{2} 3^2 - \frac{1}{2} 3 + 1 \right) = -5 \sum_{n=1}^{k} \frac{1}{27^n}$$

$$= -5 \left( -1 + \sum_{n=0}^{k} \frac{1}{27^n} \right).$$

Hence, by Example 5.4,

$$s_{3k} \to -5 \left( -1 + \frac{1}{1 - (1/27)} \right) = \frac{-5}{26}.$$

Hence, $L = -5/26$, as was to be proved.

# Chapter 6

1. (a) Assume $x_n \to L$. Then $x_{n+1} \to L$ also (Theorem 4.3). But $x_{n+1} = f(x_n)$ and $f$ is continuous (everywhere, and, in particular) at $L$, so $f(x_n) \to f(L)$. Hence $L = f(L)$.

   (b) In this case, $x_0 = a$, $x_0 = a^2$, $x_2 = a^4$, $x_3 = a^8$, leading one to expect that $x_n = a^{(2^n)}$. Let's prove this by induction: we already know $x_0 = a = a^{(2^0)}$ so the claim holds for $n = 0$. Assume it holds for $n = k$. Then $a_{k+1} = f(a_k) = (a_k)^2 = (a^{(2^k)})^2 = a^{2(2^k)} = a^{(2^{k+1})}$ so it holds for $n = k+1$. Hence, by induction, the claim holds for all $n \in \mathbb{N}$. So $(x_n)$ is a subsequence of the sequence $y_n = a^n$ (namely $x_n = y_{2^n}$). Now $y_n \to 0$ for all $a \in (-1, 1)$, so $x_n \to 0$ in this case also (Theorem 4.3). If $|a| > 1$ then $x_n$ is unbounded, and hence divergent. If $a = 1$ then $x_n = 1$ for all $n$, which converges to 1, and if $a = -1$, $x_n = 1$ for all $n \geq 1$, so again, $x_n \to 1$. Hence, the set of values of $a$ for which $(x_n)$ converges is precisely $[-1, 1]$.

   (c) $f$ is polynomial, hence continuous. If $x$ is a fixed point of $f$ then $x^2 - x + c = 0$. This equation has no (real) solutions if $c > 1/4$ (since the quadratic polynomial's discriminant is $(-1)^2 - 4c < 0$ in this case). Since $f$ is continuous and has no fixed points, no sequence generated by iterating $f$ can converge, by part (i).

   (d) An easy example is $f(x) = x/2$ for $x \neq 0$ and $f(0) = 1$, and $a = 1$. Then $x_n = 1/2^{n+1}$ (check it!) so $x_n \to 0$, and $f(x_n) = x_{n+1} \to 0 \neq 1 = f(0)$.

2. (a) $f : [1, 2] \to \mathbb{R}$ is continuous, $f(1) = -1$, $f(2) = 3$, so 0 is a number between $f(1)$ and $f(2)$. Hence, by the Intermediate Value Theorem, there exists $c \in [1, 2]$ such that $f(c) = 0$.

   (b) Using my pocket calculator, I find that

   $$f(1.5) > 0 \Rightarrow c \in [1, 1.5]$$
   $$f(1.25) < 0 \Rightarrow c \in [1.25, 1.5]$$
   $$f(1.375) < 0 \Rightarrow c \in [1.375, 1.5]$$
   $$f(1.4375) < 0 \Rightarrow c \in [1.4375, 1.5]$$
   $$f(1.453125) < 0 \Rightarrow c \in [1.453125, 1.5].$$

   At this point we can stop, because every number in $[1.453125, 1.5]$ is 1.5 to one decimal place.

   (c) We need to confine $c$ to an interval of length less than $\varepsilon = 5 \times 10^{-7}$. After $n$ iterations of the bisection method, $c$ is confined to an interval of width $1/2^n$, so we need $n$ sufficiently large that $1/2^n < \varepsilon$, or, equivalently, $2^n > 1/\varepsilon = 2 \times 10^6$. If we assume that $\ln : (0, \infty) \to \mathbb{R}$ is an increasing function (which it is), it follows that $n \ln 2 > \ln 2 + 6 \ln 10$, so we require that

$$n > 1 + 6 \frac{\ln 10}{\ln 2} \in (20, 21).$$

Hence, 21 iterations of the bisection method will be enough.

## Chapter 7

1.

| $P$ | $Q$ | $R$ | $R \Rightarrow \neg P$ | $P \Rightarrow (Q \wedge (R \Rightarrow \neg P))$ |
|---|---|---|---|---|
| 0 | 0 | 0 | | 1 |
| 0 | 0 | 1 | | 1 |
| 0 | 1 | 0 | | 1 |
| 0 | 1 | 1 | | 1 |
| 1 | 0 | 0 | | 0 |
| 1 | 0 | 1 | | 0 |
| 1 | 1 | 0 | 1 | 1 |
| 1 | 1 | 1 | 0 | 0 |

*Remark: I haven't bothered to fill in all of the columns here because I know that any statement of the form $P \Rightarrow$ (something) is true when $P$ is false (so I immediately get the top 4 entries), and if $P$ is true then $P \Rightarrow$ (something) is false when (something) is false (which gives me the next 2 entries, since $Q \wedge$ (something) is false when $Q$ is false).*

2. (a) $\forall x \in \mathbb{Q}, \exists y \in \mathbb{R} \backslash \mathbb{Q}, \exists z \in \mathbb{R} \backslash \mathbb{Q}, x = y/z$. False: $x = 0$ is rational, but is not the quotient of any pair of irrational numbers (note that $y/z = 0$ if and only if $y = 0$, which is not irrational).

   (b) $[\exists K \in \mathbb{R}, \forall n \in \mathbb{Z}^+, n + 1/n \geq K] \wedge [\forall M \in \mathbb{R}, \exists m \in \mathbb{Z}^+, m + 1/m > M]$. True: 0 is a lower bound on $A$ (since every element of $A$ is positive), and given any $M \in \mathbb{R}$, there is some positive integer $m$ such that $m > M$ (by the Archimedean Property), and then $m + 1/m \in A$ and $m + 1/m > m > M$.

   (c) $\forall (x, y) \in \mathbb{R}^2, x^3 - x = y^3 - y \Rightarrow x = y$. False: $1^3 - 1 = 0^3 - 0$ but $1 \neq 0$.

3. (a) Between any two distinct real numbers, there is a rational number. *Remark: this is the Density Theorem for $\mathbb{Q}$.*

   (b) The function $f : \mathbb{R} \to \mathbb{R}, f(x) = x^3 - x$ is surjective.

4. (a) $\neg P : \exists x \in \mathbb{R}, [x^2 > x \wedge x^4 \leq x^2]$.
   $\neg P$ is true; for example, $x = -1/2$ satisfies $x^2 > x$ and $x^4 \leq x^2$.

(b) $\neg Q : \exists x \in \mathbb{R}, \forall y \in \mathbb{R}, |y| \neq x + y$.

$\neg Q$ is true. To see this, note that $|y| \geq y$, so $|y| - y \geq 0$. Hence, there does exist a real number $x$ which is not equal to $|y| + y$ for any choice of $y$, namely $x = -1$ (for example – any negative real number will do).

(c) $\neg R : \forall x \in \mathbb{R}, \exists y \in \mathbb{R}, x + y \geq |y|$.

$R$ is true. For example, take $x = -1$. Then, for all $y \in \mathbb{R}$, $x + y < y \leq |y|$.

## Chapter 8

1. Let $f(x) = (x^3 - 8)/(x^2 - 4)$ and $(x_n)$ be any sequence in $\mathbb{R}\backslash\{-2, 2\}$ such that $x_n \to 2$. Then

$$f(x_n) = \frac{x_n^3 - 8}{x_n^2 - 4} = \frac{(x_n - 2)(x_n^2 + 2x_n + 4)}{(x_n - 2)(x_n + 2)} = \frac{x_n^2 + 2x_n + 4}{x_n + 2} \to 3$$

by the Algebra of Limits (Theorem 3.5). Hence, $\lim_{x \to 2} f(x)$ exists and equals 3.

2. (a) Let $K \in \mathbb{R}$ be given. Since $a_n \to L$, there exists $N_1 \in \mathbb{Z}^+$ such that, for all $n \geq N_1$, $|a_n - L| < 1$, and hence, $a_n < L + 1$. Let $K' = K - L - 1$. Since $b_n \to -\infty$, there exists $N_2 \in \mathbb{Z}^+$ such that, for all $n \geq N_2$, $b_n \leq K'$. Hence, for all $n \geq N = \max\{N_1, N_2\}$,

$$a_n + b_n < L + 1 + b_n \qquad \text{(since } n \geq N_1)$$
$$\leq L + 1 + K' = K \qquad \text{(since } n \geq N_2).$$

Hence $a_n + b_n \to -\infty$.

(b) No: take $a_n = n$ and $b_n = 2n$. Then $a_n \to \infty$ and $b_n \to \infty$ (by the Sweep Rule), but $a_n - b_n = -n$ which is unbounded (below), so cannot converge (to 0 or anything else).

(c) No: take $a_n = n^2$ and $b_n = n$. Then $b_n \to \infty$ and $a_n \to \infty$ (by the Sweep Rule), but $a_n/b_n = n$ which is unbounded (above), so cannot converge (to 1 or anything else).

3. Let $(x_n)$ be any sequence that diverges to infinity. We must show that $f(x_n) \to 1$. So, let $\varepsilon > 0$ be given. Then, since $x_n \to \infty$, there exists $N \in \mathbb{Z}^+$ such that, for all $n \geq N$, $x_n > \sqrt{2/\varepsilon}$. Hence, for all $n \geq N$,

$$|f(x_n) - 1| = \left| \frac{-2}{x_n^2 + 1} \right| < \frac{2}{x_n^2} < \varepsilon.$$

4. **Definition** $\lim_{x \to \infty} f(x) = \infty$ if, for every sequence $(x_n)$ such that $x_n \to \infty$, $f(x_n) \to \infty$.

Let $f(x) = x^3$ and $x_n \to \infty$. We must show that $f(x_n) \to \infty$. So, let $K \in \mathbb{R}$ be given. Since $x_n \to \infty$, there exists $N \in \mathbb{Z}^+$ such that, for all $n \geq N$, $x_n > |K| + 1$. Hence, for all $n \geq N$,
$$f(x_n) > (|K| + 1)^3 = |K|^3 + 3|K|^2 + 3|K| + 1 > 3|K| \geq K.$$

## Chapter 9

1. (a) Let $(x_n)$ be any sequence in $\mathbb{R} \setminus \{3\}$ such that $x_n \to 3$. Then
$$\frac{f(x_n) - f(3)}{x_n - 3} = \frac{x_n^2 - 9}{x_n - 3} = x_n + 3 \to 6.$$
Hence $f'(3) = 6$.

(b) Let $(x_n)$ be any sequence in $\mathbb{R} \setminus \{0, 1\}$ such that $x_n \to 1$. Then
$$\frac{f(x_n) - f(1)}{x_n - 1} = \frac{1/x_n^2 - 1}{x_n - 1} = \frac{1 - x_n^2}{x_n^2(x_n - 1)} = \frac{1 + x_n}{-x_n^2} \to -2.$$
Hence $f'(1) = -2$.

(c) Let $(x_n)$ be any sequence in $\mathbb{R} \setminus \{0\}$ such that $x_n \to 0$. Then
$$s_n = \frac{f(x_n) - f(0)}{x_n - 0} = \frac{x_n^2 \sin x_n}{x_n} = x_n \sin x_n.$$
Hence $-x_n \leq s_n \leq x_n$ for all $n$, so $s_n \to 0$ by the Squeeze Rule. Hence $f'(0) = 0$.

2. Let $g : \mathbb{R} \to \mathbb{R} \setminus \{0\}$ such that $g(x) = 1 + x^6$ and $h : \mathbb{R} \setminus \{0\} \to \mathbb{R}$ such that $h(x) = x^{-7}$. Then $g'(1) = 6(1)^5 = 6$, $g(2) = 1 + 1^6 = 2$, and $h'(2) = -7(2)^{-8} = -7/256$. Hence, by the Chain Rule,
$$f'(1) = (h \circ g)'(1) = h'(2)g'(1) = -\frac{42}{256} = -\frac{21}{128}.$$

3. Let $F = g \circ f : A \to C$. By the Chain Rule, $F$ is differentiable at $a$ with $F'(a) = g'(f(a))f'(a)$. But $h$ is differentiable at $F(a) = g(f(a))$, so by the Chain Rule again, $h \circ g \circ f = h \circ F$ is differentiable at $a$, and
$$(h \circ g \circ f)'(a) = h'(F(a))F'(a) = h'((g \circ f)(a))g'(f(a))f'(a).$$

4. Let $r : \mathbb{R} \to \mathbb{R}$ such that $r(x) = -x$. Then $r$ is certainly differentiable, and $r'(x) = -1$.

(a) Let $f$ be even. Then $f \circ r = f$, so, by the Chain Rule
$$f'(x) = (f \circ r)'(x) = f'(r(x))r'(x) = f'(-x)(-1).$$
Hence $f'$ is odd.

(b) Let $f$ be odd. Then $f \circ r = -f$, so, by the Chain Rule
$$-f'(x) = (f \circ r)'(x) = f'(r(x))r'(x) = f'(-x)(-1).$$
Hence $f'$ is even.

## Chapter 10

1. (a) Recall the definition of radius of convergence (Definition 10.5)

$$R = \sup A, \qquad \text{where} \quad A = \{|x| : \sum_{n=0}^{\infty} \left| \frac{nx^{4n}}{2^{2n+1}} \right| \text{ converges}\}.$$

Let $b_n = nx^{4n}/2^{2n+1}$. Then

$$\frac{|b_{n+1}|}{|b_n|} = \frac{(n+1)|x|^{4n+4}}{2^{2n+3}} \times \frac{2^{2n+1}}{n|x|^{4n}}$$

$$= \frac{n+1}{n}\frac{1}{4}|x|^4 = \left(1 + \frac{1}{n}\right)\frac{|x|^4}{4} \to \frac{|x|^4}{4}.$$

Hence, if $|x| < \sqrt{2}$, then $|b_{n+1}|/|b_n|$ converges to a limit less than 1, so, by the Ratio Test, the power series converges absolutely. Hence, every $|x| \in [0, \sqrt{2})$ is an element of $A$, so $R \geq \sqrt{2}$. Conversely, if $|x| > \sqrt{2}$, then $|b_{n+1}|/|b_n|$ converges to a limit greater than 1 so by the Ratio Test, the power series does *not* converge absolutely. Hence, every real number greater than or equal to $\sqrt{2}$ is an upper bound on $A$, so $R \leq \sqrt{2}$. Hence $R = \sqrt{2}$.

(b) Let $f(x) = \sum_{n=0}^{\infty} \frac{n!}{2^{n(n+1)}} x^{(n^2)}$. Consider first the series at $x = 2$,

$$f(2) = \sum_{n=0}^{\infty} \frac{n!}{2^{(n^2+n)}} 2^{(n^2)} = \sum_{n=0}^{\infty} \frac{n!}{2^n}.$$

Let $c_n = n!/2^n$. Then $c_{n+1}/c_n = (n+1)/2 \geq 1$ for all $n \geq 1$, so, for all $n \geq 2$, $c_n \geq c_1 = 1/2$. Hence $c_n \not\to 0$, so the series $\sum_{n=0}^{\infty} c_n$ does not converge (by the Divergence Test). Hence, by Theorem 10.7, $R \leq 2$ (if $R$ were greater than 2, the series would converge absolutely, and hence converge, at $x = 2$).

Now consider the series at general $x$. Let $b_n = \frac{n!}{2^{n(n+1)}} x^{(n^2)}$. Then if $x \neq 0$, $b_n \neq 0$ and

$$\frac{|b_{n+1}|}{|b_n|} = \frac{|x|}{4}(n+1)\left(\frac{|x|}{2}\right)^{2n}. \tag{T.2}$$

This sequence converges to 0 if $|x|/2 < 1$ (see Example 5.13). Hence, by the Ratio Test, for all $x \in \mathbb{R}$ with $|x| < 2$, the series converges absolutely. Hence (by the definition of radius of convergence) $R \geq 2$. Combining with the first part, we see that $R = 2$.

*Remark: alternatively, we could skip the first part (the case $x = 2$) and deduce from equation (T.2) that $|b_{n+1}|/|b_n|$ is unbounded for $|x| \geq 2$, whence it follows that $|b_n|$ is unbounded, and hence $\sum b_n$ does not converge absolutely. I think it's more elegant to argue as above: observe that $f(2)$ diverges, and deduce from Theorem 5.20 that the disk of convergence can't have radius larger than 2.*

2. Let $f(x) = \sum_{n=0}^{\infty} x^n$. This has radius of convergence 1, and converges to $f(x) = 1/(1-x)$ (Example 10.1). Hence, by Theorem 10.10, $f$ is differentiable on $(-1, 1)$ and

$$f'(x) = \sum_{n=1}^{\infty} nx^{n-1}.$$

But, by the Quotient Rule,

$$f'(x) = \frac{d}{dx} \frac{1}{1-x} = \frac{1}{(1-x)^2}.$$

Equating these two expressions for $f'(x)$ establishes the claim. Now note that $f(1/2) = 1/(1/2)^2 = 4$. Hence

$$\sum_{n=1}^{\infty} \frac{n+1}{2^n} = \frac{1}{2} \sum_{n=1}^{\infty} n(1/2)^{n-1} + \sum_{n=0}^{\infty} (1/2)^n - (1/2)^0$$

$$= \frac{1}{2} f'(1/2) + f(1/2) - 1$$

$$= 2 + 2 - 1 = 3.$$

## Chapter 11

1. (a) We can prove this by induction on $n$. The formula certainly holds for $n = 1$, since

$$1^3 = \frac{1}{4}(1)^2(2)^2.$$

Assume it holds for some $n = k \geq 1$. Then

$$\sum_{j=1}^{k+1} j^3 = \sum_{j=1}^{k} j^3 + (k+1)^3 = \frac{1}{4}k^2(k+1)^2 + (k+1)^3$$

$$= \frac{1}{4}(k+1)^2(k^2 + 4k + 4) = \frac{1}{4}(k+1)^2(k+2)^2.$$

Hence, the formula holds for $n = k + 1$. Hence, by induction, the formula holds for all $n \in \mathbb{Z}^+$.

(b) For each $n \in \mathbb{Z}^+$, consider the regular dissection of size $n$,

$$\mathscr{D}_n = \{0, \frac{1}{n}, \frac{2}{n}, \dots, 1\}.$$

Since $x^3$ is increasing on $[0, 1]$, its lower sum with respect to this dissection is

$$l_{\mathscr{D}_n} = \sum_{j=1}^{n} \frac{1}{n} \left( \frac{j-1}{n} \right)^3 = \frac{1}{n^4} \sum_{j=1}^{n} (j-1)^3 = \frac{1}{n^4} \sum_{k=0}^{n-1} k^3$$

$$= \frac{1}{4n^4} (n-1)^2 n^2 \to \frac{1}{4}$$

by part (a). Similarly, its upper sum with respect to this dissection is

$$u_{\mathscr{D}_n} = \sum_{j=1}^{n} \frac{1}{n} \left( \frac{j}{n} \right)^3 = \frac{1}{n^4} \sum_{j=1}^{n} j^3 = \frac{1}{4n^4} n^2 (n+1)^2 \to \frac{1}{4}.$$

Hence, $u_{\mathscr{D}_n} - l_{\mathscr{D}_n} \to 0$ so, by Theorem 11.12, $x^3$ is integrable on $[0, 1]$ and

$$\int_0^1 x^3 \, dx = \lim u_{\mathscr{D}_n} = \frac{1}{4}.$$

2. The piecewise linear function $f : [0, 3] \to \mathbb{R}$ depicted below has the required properties with respect to the dissections $\mathscr{D} = \{0, 3\}$, $\mathscr{D}' = \{0, 2, 3\}$.

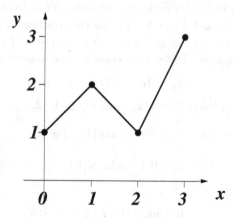

We see that

$$l_{\mathscr{D}}(f) = 3 \times 1 = 3$$
$$u_{\mathscr{D}}(f) = 3 \times 3 = 9$$
$$l_{\mathscr{D}'}(f) = 2 \times 1 + 1 \times 1 = 3$$
$$u_{\mathscr{D}'}(f) = 2 \times 2 + 1 \times 3 = 7.$$

So in this case, passing from $\mathscr{D}$ to its refinement $\mathscr{D}'$ improves the overestimate (the upper sum), but makes no change to the underestimate (the lower sum).

3. The function $f : [0, 1] \to \mathbb{R}$ defined so that

$$f(x) = \begin{cases} 1 & \text{if } x \in \mathbb{Q} \\ -1 & \text{if } x \notin \mathbb{Q} \end{cases}$$

is not Riemann integrable. To see this note that for *any* dissection $\mathscr{D} = \{a_0, \ldots, a_n\}$ of $[0, 1]$,

$$l_{\mathscr{D}}(f) = \sum_{j=1}^{n}(a_j - a_{j-1})(-1) = -\sum_{j=1}^{n}(a_j - a_{j-1}) = -1$$

since every interval $[a_{j-1}, a_j]$ contains an irrational number (Theorem 1.34), and

$$u_{\mathscr{D}}(f) = \sum_{j=1}^{n}(a_j - a_{j-1})(+1) = \sum_{j=1}^{n}(a_j - a_{j-1}) = 1$$

since every interval $[a_{j-1}, a_j]$ contains a rational number (Theorem 1.25). Hence $l(f) = \sup\{-1\} = -1$ and $u(f) = \inf\{1\} = 1$, so $l(f) \neq u(f)$. However, $f^2$ is the constant function, $f^2(x) = 1$, which is certainly integrable (since it's continuous, for example).

4. We first prove the claim in the case where $g$ differs from $f$ only at a single point $c \in [a, b]$. Let $\delta : [a, b] \to \mathbb{R}$ be the function $\delta(x) = g(x) - f(x)$. Then $\delta(x) = 0$ for all $x \neq c$ and $\delta(c) = g(c) - f(c) = K \neq 0$. I claim that $\delta$ is integrable. To see this, consider the sequence of dissections

$$\mathscr{D}_n = \{a, c - 1/n, c + 1/n, b\}.$$

The upper and lower Riemann sums for each $\mathscr{D}_n$ are

$$u_{\mathscr{D}_n}(\delta) = (c - 1/n - a) \times 0 + \frac{2}{n}\max\{K, 0\} + (b - c - 1/n) \times 0 \leq \frac{2}{n}|K|$$

$$l_{\mathscr{D}_n}(\delta) = (c - 1/n - a) \times 0 + \frac{2}{n}\min\{K, 0\} + (b - c - 1/n) \times 0 \geq -\frac{2}{n}|K|$$

so

$$0 \leq u_{\mathscr{D}_n}(\delta) - l_{\mathscr{D}_n}(\delta) \leq \frac{4}{n}|K|$$

and hence $u_{\mathscr{D}_n}(\delta) - l_{\mathscr{D}_n}(\delta) \to 0$ (Squeeze Rule). Hence $\delta$ is Riemann integrable by Theorem 11.12. But then $g = f + \delta$ is Riemann integrable by Theorem 11.21. (Strictly speaking, the argument above works only if $c \in (a, b)$. If $c = a$, we should instead use the sequence of dissections $\mathscr{D}_n = \{a, a + 1/n, b\}$, and if $c = b$, $\mathscr{D}_n = \{a, b - 1/n, b\}$.)

Now consider the case where $g$ differs from $f$ only at $c_1, c_2, \ldots, c_p \in [a, b]$. Then we can define a finite chain of functions $g_0, \ldots, g_p$ as follows: $g_0 = f$ and, for each $j = 1, \ldots, p$,

$$g_j(x) = \begin{cases} g_{j-1}(x) & \text{if } x \neq c_j \\ g(x) & \text{if } x = c_j \end{cases}.$$

By construction, $g_0 = f$, $g_p = g$, and $g_j$ differs from $g_{j-1}$ at only one point. Hence, applying the result just proved $p$ times, we deduce that $g$ is Riemann integrable (since $g_0$ is integrable, so is $g_1$; since $g_1$ is integrable, so is $g_2$ etc.).

## Chapter 12

1. By Definition 12.7, $f(x) = \exp(x \log b)$, which is a composition of the smooth functions $x \mapsto \exp(x)$ and $x \mapsto x \log b$, so is smooth, by the Chain Rule. Further,

$$f'(x) = \log b \exp(x \log b) > 0$$

for all $x$ since $b > 1$ (so $\log b > \log 1 = 0$). Hence $f$ is strictly increasing and injective (Proposition 9.24). Let $y \in (0, \infty)$. Since exp is surjective (Proposition 10.14), there exists $X \in \mathbb{R}$ such that $\exp(X) = y$. But then $f(X/\log b) = \exp(X) = y$. Hence $f$ is surjective.

(a) Let $x \in \mathbb{R}$ and $y \in (0, \infty)$ be related by $y = f(x)$. Then

$$y = \exp(x \log b)$$
$$\Rightarrow \quad \log y = \log(\exp(x \log b)) = x \log b$$
$$\Rightarrow \quad \frac{\log y}{\log b} = x.$$

Hence $\log_b y = \log y / \log b$ for all $y \in (0, \infty)$.

(b) $\log_b(xy) = \dfrac{\log(xy)}{\log b} = \dfrac{\log x + \log y}{\log b} = \log_b x + \log_b y$
   by Proposition 12.2.

(c) $f(\log_b(x^y)) = x^y$ by definition of $\log_b$, and

$$f(y \log_b x) = \exp(y \log_b x \log b) = \exp(y \log x) = x^y$$

by definition of $x^y$. Since $f(\log_b(x^y)) = f(y \log_b x)$ and $f$ is injective, it follows that $\log_b(x^y) = y \log_b x$.

(d) Follows immediately from part (a) and Proposition 12.4.

$\log_b 1 = \log 1/\log b = 0$, $\log_b b = \log b/\log b = 1$, $\log_b e = \log e/\log_b = 1/\log b$.

2. Consider the function $f : (0, \infty) \to \mathbb{R}$, $f(x) = \frac{\log x}{x}$. This is differentiable, and

$$f'(x) = -\frac{1}{x^2}(\log x - 1) < 0$$

for all $x > e$. Hence the function is decreasing for all $x > e$. Now $x_n = f(n)$, so this sequence is decreasing for $n > e$. For all $n \geq 1$, $\log n \geq 0$, so $(x_n)$ is bounded below. Hence, by the Monotone Convergence Theorem, $x_n$ converges to some limit $L$, say. Consider the subsequence $x_{n^2}$. This also converges to $L$ (Theorem 4.3). But

$$x_{n^2} = \frac{2 \log n}{n^2} = \frac{2}{n} x_n \to 0 \times L = 0$$

by the Algebra of Limits. Hence $L = 0$ (limits are unique).

Now $y_n = \exp(x_n)$, and exp is continuous, so $y_n \to \exp(0) = 1$, since $x_n \to 0$.

3. Let $s_k = \sum_{n=2}^{k} \frac{1}{n \log n}$. We will show that the sequence $s_k$ is unbounded above (and so does not converge). For each $k \geq 3$, consider the function $f : [2, k+1] \to \mathbb{R}$, $f(x) = \frac{1}{x \log x}$. This is differentiable, hence continuous, hence Riemann integrable. So $\int_2^{k+1} f$ exists, and is bounded above by the upper Riemann sum $u_{\mathscr{D}}(f)$, where $\mathscr{D} = \{2, 3, \ldots, k+1\}$ is the regular dissection of $[2, k+1]$ of size $k-1$. Since both $x$ and $\log x$ are increasing functions of $x$, $f(x)$ is decreasing, and so

$$u_{\mathscr{D}}(f) = \sum_{j=2}^{k} (1) \times f(j) = \sum_{j=2}^{k} \frac{1}{j \log j} = s_k.$$

Hence,

$$s_k = u_{\mathscr{D}} \geq \int_2^{k+1} f.$$

It is straightforward to compute $\int_2^{k+1} f$ using version 2 of the Fundamental Theorem of the Calculus. One simply notes that

$$F(x) = \log(\log x)$$

is an antiderivative of $f$, so

$$\int_2^{k+1} f = F(k+1) - F(2) = \log(\log(k+1)) - \log(\log 2).$$

The sequence $\log(\log(k+1))$ is unbounded above, so $s_k$ is unbounded above.

*Remark: this trick of interpreting $\sum_{n=1}^{k} f(n)$ as the upper (or lower) Riemann sum of $f$ on a suitable interval is widely applicable and can be used to prove both convergence or (as in this example) divergence of the series. Of course, it's only useful if you can evaluate $\int_{1}^{k} f$ more easily than $s_k$. This is often the case, because version 2 of the Fundamental Theorem of the Calculus is a powerful tool for computing integrals.*

# Index

Printed in the United States
By Bookmasters